BestMasters

Springer awards „BestMasters" to the best master's theses which have been completed at renowned universities in Germany, Austria, and Switzerland.

The studies received highest marks and were recommended for publication by supervisors. They address current issues from various fields of research in natural sciences, psychology, technology, and economics.

The series addresses practitioners as well as scientists and, in particular, offers guidance for early stage researchers.

Jonathan Orlando Zaddach

Climate Policy Under Intergenerational Discounting

An Application of the Dynamic Integrated Climate-Economy Model

Springer Gabler

Jonathan Orlando Zaddach
Nürnberg, Germany

Master Thesis Albert-Ludwigs-University of Freiburg, 2015 u.d.t.: "Climate Policy Under Intergenerational Discounting: An Application of DICE."

BestMasters
ISBN 978-3-658-12133-4   ISBN 978-3-658-12134-1 (eBook)
DOI 10.1007/978-3-658-12134-1

Library of Congress Control Number: 2015956103

Springer Gabler
© Springer Fachmedien Wiesbaden 2016

Printed on acid-free paper

Springer Gabler is a brand of Springer Fachmedien Wiesbaden
Springer Fachmedien Wiesbaden is part of Springer Science+Business Media
(www.springer.com)

Climate Policy Under Intergenerational Discounting: An Application of DICE

Abstract: Both *consumer sovereignty* and *intergenerational equity* are eligible stances in the discussion on how to weigh future generations against present ones. At best, they should not be played off against each other. Under the representative agent approach, however, one has to take sides, which differ enormously in their consequences for climate policy. Krysiak (2010) shows that this dilemma can be overcome in the framework of an overlapping generations (OLG) model. For a set of normative assumptions, he derives a utility discounting scheme which assigns to all generations equal weights without disregarding their true preferences. In this thesis, we apply the discounting scheme in the latest DICE model and present its implications for optimal climate policy. Furthermore, we carry out an OFAT sensitivity analysis to check the discounting scheme for robustness. It turns out that the proposed discounting scheme fails in incorporating *consumer sovereignty* and *intergenerational equity* sufficiently. Krysiak's approach should deliver conceptual guidance for future research, though.

Table of Contents

1. Introduction

Cost-benefit-analysis is an indispensable instrument for evaluating the long-term impacts of the efficiency of policy measures. To make costs and benefits which occur at different points in time comparable, economists determine their present values through the application of a discount rate. In many cases, the size of this rate determines whether a project is classified as attractive or unattractive. This is particularly true for projects with long time horizons, such as the abatement of greenhouse gas (GHG) emissions. [1] Most commonly, economic theory derives this discount rate in the framework of optimal growth models in the tradition of Frank Ramsey.[2] The result, the well-known Ramsey formula, states that the consumption discount rate should be equal to the sum of the pure rate of time preference and the product of the elasticity of the marginal utility of consumption and the growth rate of consumption.

In the context of climate change, the correct parametrization of the Ramsey formula is very important as slightly different rates entail entirely different climate policy outcomes. With regard to the pure rate of time preference, this becomes especially difficult because different values can be motivated by different normative concepts. Not surprisingly, the question of its correct parametrization has given rise to a long and intensive discussion in climate economics. This debate, sometimes referred to as the Stern-Nordhaus Debate, as it took place most prominently between the two economists William Nordhaus and Nicholas Stern, can be characterized by two distinct lines of arguments:[3]

Nordhaus (2007) argues for the so-called *consumer sovereignty* approach, which demands that public projects should be evaluated with a discount rate that is compatible with observed time preferences. These are claimed to be given by the real interest rate, as it represents the opportunity costs of private investments. Because public and private investment should be evaluated with the same standards any discount rate below or above the real interest rate would induce inefficient investment decisions. This position is held, for example, by Stigler and Becker (1977), Samuelson and Nordhaus (1989), Manne et al. (1995) and Nordhaus (2008).

---

[1] See Arrow et al. (2013), p. 1.
[2] See Bayer (2000), p. 1.
[3] William Nordhaus and Nicholas Stern were not the only ones involved in this debate. For example, Weitzman (2007), Gollier (2006), Gollier (2010), Dasgupta (2007, 2008) or Tol and Yohe (2006) also made important contributions to it.

In contrast, Stern (2007) argues that in long-term decision-making problems, actions taken at present will not only affect the well-being of individuals who live today, but also of those who are not yet in existence. In this regard, there is a distributional problem between individuals, not an intertemporal allocation problem of a single individual. From Stern's perspective, it would be ethically inappropriate to discriminate between individuals purely on the bases of their time of birth. The logical consequence is to set $\rho = 0$. This argument is called the *intergenerational equity* approach and is in the tradition of Pigou (1920), Ramsey (1928), Harrod (1948) and Solow (1974) or, in the context of the recent climate change debate Cline (1992).

As both arguments have their merits, it would be desirable to have a discounting scheme that incorporates both. Krysiak (2010) shows that this can be achieved in an overlapping generations (OLG) model which has been discussed by a number of authors who examine the relation between representative agent (RA) models and overlapping generations models. For example, Stephan et al. (1997) and Howarth (1998, 2000) analyze under which conditions representative agent models can be calibrated to yield the same outcome as OLG models. They employ numerical simulations of integrated assessment models. Schneider et al. (2008, 2012), in contrast, derive the relation between these two frameworks in a continuous time set-up and identify several shortcomings of the RA approach. Krysiak's approach differs from these studies in so far as he derives the discounting scheme from a set of normative assumption.

However, Krysiak (2010) does not draw any policy conclusions. In order to fill this gap, we apply the proposed discounting scheme in the latest version of the DICE model and present its implications for an optimal climate policy. It calls for a carbon price path which is just between the ones proposed by Nordhaus and Stern. Following this path, industrial emissions should be stabilized by mid-century and subsequently be substantially reduced. In this scenario, average global temperature will increase up to 2.5°C above the preindustrial level by the end of this century. We also analyze to what extent the choice of all relevant parameters drives the outcome of the intertemporal welfare maximization. It turns out that in the short run time preferences dominate the discounting scheme, whereas in the long-run only the risk of extinction affects the optimal growth path. This casts a cloud over the proposed discounting scheme as it seems to incorporate the two ethical stances successively instead of simultaneously.

This thesis is organized as follows: In section 2, we introduce the conceptual framework of Integrated Assessment Models (IAMs) and describe the specific characteristics of the model which will be applied here, DICE-2013R. In section 3, we give an overview of the Stern-Nordhaus debate and present the conceptual differences between the *consumer sovereignty* approach and the *intergenerational equity* approach. Section 4 introduces the concept of intergenerational discounting and explains how it is derived from overlapping generations models. In section 5, we firstly derive the discounting scheme as proposed by Krysiak (2010) (5.1), then present its main implications for optimal climate policy (5.2) and finally carry out a sensitivity analysis to check our results for robustness (5.3). The conclusions are presented in Section 6.

2. DICE-2013R and Other Integrated Assessment Models

"Integrated Assessment Models" (IAMs) are computer simulation models that integrate insights from different disciplines such as ecology, earth sciences and economics.[4] According to Weyant et al. (1996) IAMs serve three purposes: First, they allow assessing climate change control policies. Second, they integrate the different dimensions of climate change in the same conceptual framework. Third, they help to quantify the relative importance of global warming within the limits of other environmental and non-environmental problems that are faced by mankind.

For the climate issue there are more than 50 IAMs that differ with respect to modeling structure, complexity and assumptions regarding society parameters and the climate system.[5] IAMs can be divided into two different types: policy evaluation models and policy optimization models. Policy evaluation models are usually recursive or equilibrium models that simulate the effect of a single policy option on the biosphere, the climate and the economy.[6] In contrast, policy optimization models attempt either to determine the optimal policy or to simulate the impact of an efficient level of carbon abatement on the global economy.[7] Optimal solutions are determined by maximizing an objective function or welfare function that are characterized by either regulatory efficiency, where expected costs and benefits of climate protection are traded off against each other , or regulatory cost-effectiveness - a solution which minimizes the costs of achieving a particular goal.[8]

IAMs that calculate dynamically optimal emission paths have several specific features in common: They all postulate a single long-lived representative individual whose preferences provide the basis for the optimization. Furthermore, abatement costs and climate damages must be expressed in a common unit and the aggregated climate damage function is represented by a simple power function of temperature change. Last, to compare the costs over long-time horizons, a discount rate is applied.[9]

IAMs are frequently used in the field of climate economics as they allow to break down the complexity of the economic, climate and social systems to a very basic structure and to model their interdependencies over time in a consistent framework. Proponents of

---

[4] See Jeroen P van der Sluijs (2002), p. 1.
[5] See ibidem, p. 2.
[6] See Kelly and Kolstad (1999), p. 4 and Nordhaus and Sztorc (2013), p. 5.
[7] See Kelly and Kolstad (1999), p. 4.
[8] See Kelly and Kolstad (1999), p. 4 and Nordhaus and Sztorc (2013), p. 5.
[9] See Parson and Fisher-Vanden (1997), p. 605ff.

climate modeling do not claim that IAMs provide "definitive answers" to climate change related questions but rather consider them as helpful tools to understand how changes in one system affect changes in another system. Even if outcomes might not necessarily be correct, they can "*at least* [give] *internally consistent* [answers] *and at best provide a state-of-the-art description of the impacts of different forces and policies.*"[10]

Nevertheless, the value of climate policy derived from integrated assessment optimization is controversial and was strongly challenged only recently. One central point of criticism is that current models underestimate substantial risks of climate change.[11] For instance, the assumption that climate damage can be represented by a simple power function is thought to be quite implausible as this means that damage is still modest even if it exceeds some apparently highly dangerous thresholds.[12,13] The omission of key factors such as large-scale migrations,[14] the potentially irreversible nature of climate damage[15] or (possible) "tipping points"[16] is another weakness of integrated assessment optimization.

Most of these problems in modeling arise due to uncertainties regarding the climate system, the ecosystem, the economy and society: Existing evidence is inconclusive of how increasing GHG emissions will affect the climate system once certain thresholds are exceeded. It is also unclear what consequences this might have for the ecosystem and human well-being. Though such uncertainty issues can be mitigated, for example via Monte-Carlo-Simulations,[17] the reach of such methods is limited as uncertainty is structural: Neither do we know how strong certain factors (climate sensitivity, long-term economic growth etc.) are nor how they interrelate with each other.[18] Economists such as Weitzman (2009) have attempted to address this problem by means of different stochastic approaches. Nevertheless, there are limits to the exact and precise modeling of uncertainty.

---

[10] See Nordhaus (2008), p. 9.
[11] See Stern (2013), p. 838.
[12] See ibidem, p. 848.
[13] Using DICE-2007 Ackerman et al. (2010) show that an increase of the global temperature of up to 19°C above current average temperature implies a reduction of output of not more than 50 percent. This clearly reveals the limits of IAMs as a corresponding environment should make live on earth almost impossible.
[14] See Stern (2013), pp. 844–845.
[15] See ibidem, p. 846.
[16] The term "tipping points" refers to critical thresholds of the earth's climate system. Once these thresholds are exceeded this might cause "*abrupt climatic changes*" with "*large and potentially serious economic and ecological impacts*", see Alley et al. (2003), p. 2005.
[17] See Annan (2001), p. 270.
[18] See Weitzman (2009), p. 2.

To overcome these shortcomings, Stern (2013) identifies several key areas that require further research: First, it is important to find out if certain tipping points can be identified in the development path of the climate system. Because climate models predict that without further climate protection the median temperature is likely to exceed a threshold of 4°C one should also describe the economic and climatic consequences of such a scenario.[19] Second, IAMs should incorporate damage functions that take into account that damages from climate change do not only have short-term effects but also long-term effects on capital, land and productivity. Most current IAMs, such as DICE-2013R, do not incorporate these long-term effects and results are likely to be incorrect.[20] Last, future models need to reflect the risk of large-scale migration. It is reasonable to assume that strong changes in the climatic conditions, such as an increase of median temperature by 4°C, will cause considerable migration movements between nations and continents. History indicates that such movements involve high conflict potential and come at great costs.[21]

Apart from that, there are natural limits to modeling the climate and economic system. IAMs help to understand the complex nature of these systems and how they interrelate, but they are not capable of explaining and modeling them to the full extent.[22] Therefore, climate policy cannot fully rely on the outcome of one single model. It should rather be based on various models with different insights. Because we know that IAMs do not tell the truth, there is a need to explore additional indicators for good climate policy.[23]

DICE-2013R

One of the most popular IAMs for the cost-benefit analysis is the Dynamic Integrated Model of Climate and the Economy (DICE), which was designed by William D. Nordhaus. Since its first version from 1979, several updates with structural changes and data updates have been presented. The current version, DICE-2013R, was released in autumn 2013 and is consistent with the Fifth Assessment Report of Intergovernmental Panel on Climate Change (IPCC), released in 2013.[24] There are several other IAMs such

---

[19] See Stern (2013), p. 840.
[20] See ibidem, pp. 849–850.
[21] See ibidem, p. 1.
[22] See Nordhaus (2008), p. 9.
[23] See Stern (2013), p. 852.
[24] See Nordhaus and Sztorc (2013), p. 22.

as PAGE, FUND or MERGE that are of similar importance for the scientific community.[25] However, as DICE-2013R is state-of-the-art it should serve our purpose.[26]

The DICE model considers climate change from the perspective of neoclassical growth theory. This standard approach based on Solow (1970) assumes that there is a trade-off between today's and future consumption: If one wants to increase future consumption, today's consumption must be reduced to increase investment in capital, education or technologies. The DICE model adapts this approach in the sense that the climate system is regarded as an additional input factor: Production is positively correlated with GHG concentrations which enter as negative natural capital. The reduction of emissions comes at the cost of today's consumption. Simultaneously this reduces the damage to production caused by climate change and raises future consumption levels.[27,28]

Optimal climate policy is determined by the equilibrium in which a utilitarian social welfare function is maximized. This function ranks different consumption paths according to the preferences of a representative agent. It increases with per capita consumption $c(t)$ and with the number of existing people $L(t)$. Individual preferences are assumed to be identical and can be expressed by a constant intertemporal elasticity of substitution (CIES) utility function:[29]

Equation (1)

$$U[c(t), L(t)] = L(t) \left[\frac{c(t)^{1-\alpha}}{1-\alpha} \right]$$

The social welfare function, which is the sum of discounted welfare in all periods, is given by Equation (2):

Equation (2)

$$W[c(t), L(t)] = \sum_{t=1}^{T\,max} \frac{1}{(1+p)^t} L(t) \left[\frac{c(t)^{1-\alpha}}{1-\alpha} \right]$$

Generations are weighted in two dimensions: First, the generation's importance increases with the number of people that live in period t and with their per capita

---

[25] See Stanton et al. (2009), p. 167ff.
[26] The DICE-2013R model can be downloaded from Nordhaus' website: http://www.econ.yale.edu/~nordhaus/homepage/w
[27] See Nordhaus and Sztorc (2013), p. 4.
[28] For a detailed description of DICE-2013R see ibidem.
[29] See ibidem, p. 7.

consumption. Second, generations are weighted with regard to their time of birth, their relative importance being influenced by the pure rate of social time preference ρ and the elasticity of the marginal utility of consumption η.[30]

In the framework of Ramsey (1928)[31] this leads to the well-known Ramsey formula as a first-order condition

$$r = \frac{\partial Y}{\partial K} = \delta = \rho + \eta g\,{}^{32}$$

where the marginal opportunity cost rate r is equal to the marginal time preference rate δ.[33] The marginal time preference rate, in turn, is given by the sum of two components: The pure rate of time preference and the product of the elasticity of the marginal utility of consumption and the growth rate of consumption.

The Ramsey formula reflects two motives of discounting: On the one hand, consumption is discounted because individuals show preferences regarding the time of consumption. They rather consume today than tomorrow. This "impatience" motive is reflected by the pure rate of time preference ρ. On the other hand, consumption is discounted because future generations are likely to enjoy higher consumption levels than today's generations.[34] As the utility function shows diminishing marginal utility $\left(\frac{\partial^2 U[c(t),L(t)]}{\partial c(t)^2} < 0\right)$ future generations' marginal utility will be below the one of the earlier born generations. Therefore, redistribution in favor of the earlier born generation should increase aggregated welfare. This discounting motive is expressed by the elasticity of marginal utility η. It describes how fast the marginal utility declines as consumption increases. Higher values of η imply that the marginal utility of consumption declines more rapidly when consumption increases. It can also be interpreted as a measure of the aversion of society to inequality: The higher the elasticity of the marginal utility of consumption, the more weight is assigned to relatively poorer generations.[35]

---

[30] See Nordhaus and Sztorc (2013), p. 6.

[31] It is important to note that the derivation of the Ramsey formula is based on several assumptions: First, the economy is a competitive market, and the observed real consumption interest rate is equal to the marginal productivity of capital net of the rate of depreciation. Second, society can be represented by an infinitely-lived consumer who maximizes her utility function, see Roemer (2011), p. 372.

[32] For a detailed derivation see Appendix C.

[33] See Bayer (2003), p. 135.

[34] It is generally postulated that $g > 0$.

[35] See Nordhaus (1997), p. 316ff.

3. Consumer Sovereignty vs. Intergenerational Equity: An Overview of the Stern-Nordhaus Debate

At first glance, the Ramsey formula delivers a simple framework to the issue of discounting. However, in the context of projects with long time horizons such as the reduction of GHG emissions, the correct parametrization is indeed challenging. Here, supposedly small differences in the formula values have major impacts on policy outcomes. In particular, the value assignment for the pure rate of time preference is very controversial as different values can be motivated by different normative concepts. As mentioned above, this has caused a long and intense discussion in economics between Nicholas Stern and William Nordhaus. Most arguments concerning the correct value for ρ can be reduced to the ethical positions of *consumer sovereignty* and *intergenerational equity*. The former position states that social preferences are reflected by market outcomes, whereas the latter position argues that the parametrization of the discount rate should be based on normative decisions.[36] In the following, we present the two positions in more detail.

According to the *consumer sovereignty* position discount rates should correspond to the individual's real and observable preferences. Proponents of this approach claim that under the assumption of perfect capital markets these can be retrieved from the real rate of return of long-term investment projects.[37] As private companies base the evaluation of their investments on the real interest rate, this rate should reflect the real individual's opportunity costs of an investment. This holds for both private and public projects and therefore any deviation from this rate will entail efficiency losses.[38] In this regard, a "correct" value assignment to ρ is not as important as finding a combination of parameters $(\rho + \eta g)$ which is compatible with the real interest rate. Nordhaus (2007), for example, argues for a pure rate of time preference of $\rho = 1.5$ as this leaves us with a discount rate which is very close to the 6 percent real interest rate, he postulates.[39] The

---

[36] Because the parametrization of *consumer sovereignty* is based on the observation of people's real decision making it is also called the "descriptive approach". *Intergenerational equity*, in contrast, is referred to as the "normative approach" as the choice of parameters is exclusively based on ethical considerations, see Arrow et al. (1996), p. 131.

[37] See Paqué (2008), p. 275.

[38] See Krysiak (2010), p. 2.

[39] Along with $\eta = 2$ and $g = 2$, we obtain for $r = \rho + \eta g = 1.5 + 2 \cdot 2 = 5.5$.

position of *consumer sovereignty* has a long tradition in economics and is, for example, held by Manne et al. (1995) or Nordhaus (2008).[40]

Opponents of the *consumer sovereignty* approach present several arguments against this position. To begin with, they generally reject the assumption of perfect capital markets. This assumption is seen to be very unrealistic; after all, in the words of Stern, global warming *"is the biggest market failure the world has ever seen."*[41] Because climate protection suffers from a classical "common-pool problem"- there is no excludability and no rivalry in – long-term climate damages are systematically underestimated. This means that the social return of climate protection measures is likely to be above the private return of conventional investments and an application of the real interest rate should underestimate the benefit of climate protection.[42]

Even if capital markets were perfect, it is not certain that market outcomes coincide with socially preferred outcomes. Depending on the initial income of market participants, market processes might generate efficient results. These, however, must not necessarily coincide with what is regarded as socially optimal as market processes produce efficient but not fair outcomes. Because the amount of initial income determines the influence of specific groups on the outcome of the market process, it is plausible to assume that the interests of today's poor countries and nations, which are likely to suffer most from climate change, will have a disproportionately small impact on the intertemporal distribution of the climate good.[43]

Last and most importantly, it is argued that in long-term decision-making problems the logic of the representative agent approach does not hold since today's actions do not only affect present but also future generations. Optimal climate policy is not an intertemporal allocation problem of a single individual (as it postulated under the *consumer sovereignty* approach) but of many individuals. This position is called the *intergenerational equity* approach and is held by Stern (2007). From his perspective, it is ethically inappropriate to discriminate between individuals purely on the ground of their time of birth. Because any rate of time preference above zero implies that future generations are given less weight than present generations, Stern argues for a social time

---

[40] It is not entirely clear which interest rate should serve as a reference rate as there are various rates one could adduct. For example, the real interest rate of save treasury bonds or the average rate of fixed investments might be considered, which provides a broad range of reference values between 1,5 percent and 6-7 percent. Theoretically, even values of 26 percent could acceptable, see Buchholz and Schumacher (2009), p. 15 and Nordhaus (2008), p. 170.

[41] See Stern (2007), p. xviii.

[42] See Paqué (2008), p. 39 and Dasgupta (2008), p. 158.

[43] See Buchholz and Schumacher (2009), p. 16.

preference of almost zero ($\rho = 0,01$). Only the possibility that mankind might be eradicated due to an exogenous danger such as a meteor strike or a nuclear war legitimates a value setting of ρ above 0. To allow for a 10% chance that mankind will be eradicated within the next century, Sterns sets the social time preference rate to $\rho = 0.1$ percent.

The *intergenerational equity* approach is in the tradition of Pigou (1920), Ramsey (1928), Harrod (1948) and Solow (1974) and is based on the philosophical argument that every state has the ethical duty to give equal weights to the interests of today's and future generations. Motives as impatience or egoism, the only motives that could produce any value of $\rho > 0$, should not enter public decision problems. This argument receives philosophical support from the concept of the "veil of ignorance" as enunciated by Rawls (1972) in "A Theory of Justice". According to Rawls collective decisions should be made behind a "veil of ignorance", behind which affected people do not have any knowledge regarding their social and economic position (inter- as intratemporal). The only fair solution of this thought experiment is to assign for the time preference rate a value which gives everyone the same weight, that is $\rho = 0$.[44]

But, from the perspective of Stern's opponents, the logic of the *intergenerational equity* approach is "*not as conclusive as it claims.*"[45] First, Stern's normative stance, that all individuals should be treated equally, is not necessarily the only ethical position one could hold. For example, it could be argued that each generation should at least leave behind the social capital (material, natural, human and technological) that it was endowed with in the first place. Alternatively, social states could be ranked by a Rawlsian welfare function that maximizes the welfare of the poorest generation. Both approaches would call for completely different values for the pure rate of time preference: Whereas the former position still allows for a broad range of ρ the latter position would demand considerably higher values for ρ as present consumption should be increased to compensate for the growing productivity of future generations.[46]

Stern's critics also refer to the time dimension of the *intergenerational equity* approach which might be contrary to a reasonable and fair balance of interests. Under the concept of *intergenerational equity* the well-being of unborn individuals receives a disproportionately large weight under present policy decisions as there will be so many more individuals to come than those who live today. To demonstrate what absurd

---

[44] See Paqué (2008), p. 278.
[45] See Nordhaus (2006), p. 9.
[46] See Paqué (2008), p. 279.

consequences such a normative setting could have Nordhaus (2007) proposes the following thought experiment: In the so-called "wrinkle-experiment" it is assumed that researchers discover a wrinkle in the climate system. This wrinkle causes a tiny (0,1 percent), but ever-lasting cut-back in annual global consumption beginning in the year 2200 that can be remedied if present society is willing to invest a sufficiently large amount of income. But how much should society be willing to invest? For $\rho \approx 0$, it could be justified to reduce today's consumption by about 56 percent. As such outcomes cannot be in the interest of society one should also reject any normative concept implying it. Furthermore, this shows that the parametrization of discounting rates cannot exclusively be based on theoretical assumptions, but needs to take into account its consequences in the first place.

Last, critics such as Weitzman (2007) and Dasgupta (2008) refer to the ethical consequences of very high savings rates. These rates result from a combination of low values of the pure rate of time preference and the elasticity of the marginal utility of consumption. Since the work of Ramsey (1928), it is a standard result of economic growth theory that the optimal savings rate s* for a discrete time is given by:

$$ s^* = \frac{(1+r)^{-\frac{(\eta-1)}{\eta}}}{(1+\rho)^{\frac{1}{\eta}}} \quad 47 $$

This shows that very small values of the pure rate of time preference rate can imply extremely high savings rates. For example, if one sets $\rho = 0$ the optimal savings rate is given by $s^* = \frac{1}{\eta}$. In combination with a value of $\eta = 1$, as it is proposed by the Stern Review, the savings rate is very close to 100 percent. Thus, it could be concluded that today's generations should save almost their entire income to increase the consumption possibilities of future generations. As this result seems to be neither fair nor politically feasible Stern's critics refuse the overall choice of parameters.[48]

Nevertheless, this must not necessarily mean that the *intergenerational equity* approach has to contradict with reasonable savings behavior. Dasgupta (2008), for instance, follows Stern's argument for $\rho = 0$ and suggests solving the problem by using higher values for the elasticity of the marginal utility of consumption. Referring to an empirical

---

[47] Note that this is only the case if technology is time-invariant and linear, see Buchholz and Schumacher (2009), p. 7.
[48] See Buchholz and Schumacher (2009), p. 7.

study of Hall (1988) he argues that η should be somewhere in the range of 2 and 4 percent as this implies more reasonable savings rates between 49 and 24 percent.[49]

For optimal climate policy it makes a big difference which ethical stances one incorporates. Whereas Stern (2007) recommends immediate and drastic action, according to Nordhaus (2007) one should only pursue moderate climate protection efforts in terms of a "climate-policy-ramp". [50] Under the climate-policy-ramp investments in climate capital are postponed in favor of investment in the conventional capital stock. This increases the output and causes a productivity rise which allows investing massively in climate protection once a considerably higher level of productivity is reached. Climate policies concerned with optimal carbon taxes differ substantially between Nordhaus (2008) and Stern (2007) as the difference between them is given by an order of magnitude, as a matter of fact.[51] Most importantly, this difference is almost fully explained by the different assumptions regarding the parametrization of the social discount rate.[52]

Apparently, the question of the correct parametrization of the pure rate of time preference is key in climate economics. As both *intergenerational equity* and *consumer sovereignty* have their merits, but suffer from serious shortcomings we need a discount rate that incorporates both concepts and overcomes their respective drawbacks. But before we can indicate a potential framework for solving this dilemma, it becomes necessary to carve out the conceptual differences of the two ethical stances: As described above, the *consumer sovereignty* approach is based on the assumption that society can be represented by a single infinitely-lived individual. This individual maximizes her lifetime consumption path under the standard constraints of neoclassical growth theory. Individual time preferences enter with their full weights and the well-being of future generations is discounted simply because of their time of birth. Under the *intergenerational equity* approach, in contrast, all generations are given the same weight; there is no discounting due to the time of birth. Although conceptually we stay within the framework of the representative agent model this makes a big difference. Because in this case the existence of time preferences is disregarded one could think of a society that can be represented by a succession of individuals with a time-span of a single period.

---

[49] See Dasgupta (2007), p. 155.
[50] See Nordhaus (2007), p. 687.
[51] See Schneider et al. (2012), p. 1621.
[52] See Nordhaus (2007), p. 697ff.

Both approaches model society inaccurately as there is not just one individual generation but many and individuals' lives last beyond one period. An ideal discounting scheme must treat all individuals equally, but allows every individual the right to discount its future well-being according to its preferences, regardless of her time of birth. Such a discount rate can be derived from the framework of overlapping generations models which shall be introduced next.

4. Intergenerational Discounting and Overlapping Generations Models

The concept of "intergenerational discounting" refers to the idea that one needs to distinguish between the evaluation of projects that only affect the well-being of one generation and projects that affect the well-being of more than one generation. A discount rate that applies the traditional sort of time preferences in the context of intergenerational project evaluation does not adequately represent the overall (intertemporal) society's preferences and therefore it is insufficient.[53] To overcome this problem some authors[54] suggest that one should explicitly model the demographic structure of society which is generally done in the framework of "overlapping generations (OLG) models". These models explicitly take into account that economies are populated by successive cohorts of people with finite life-spans.[55] This implies an artificial dichotomy of equity issues: First, there is the question of the optimal allocation of consumption within the life of a generation. This is called the intragenerational effect. Second, there are the equity issues that extend beyond the life-span of a generation, which are called intergenerational effects.[56] Thus, in the framework of OLG models, we need to distinguish between the private rate of time preference and the social rate of time preference. The former characterizes the individual preference to shift consumption from the future to the present and should be exclusively applied to the evaluation of consumption streams within a generation; the latter corresponds to the rate of time preference that is socially optimal and must be applied to the evaluation of intergenerational equity issues. OLG models make a clear distinction between intergenerational equity and intergenerational efficiency.[57] They are more realistic not only because they model society's demographic structure more accurately, but also because they abstain from the assumption of an infinitely-lived agent who "*acts as a trustee on behalf of all present and future generations.*"[58] In an OLG model, individuals do not need to exhibit altruistic preferences; here agents save during working years and consume all their savings during retirement.[59] The first OLG model was developed by Samuelson in 1958 during his work on the determination of interest rates. Since then

---

[53] See Schelling (1999), p. 99ff.
[54] See for example Bayer (2003).
[55] See Howarth (1998), p. 135.
[56] See Bayer (2004), p. 150.
[57] See Stephan et al. (1997), p. 27.
[58] See ibidem, p. 28.
[59] See ibidem, p. 27.

OLG models have been applied to several areas of economic research such as microeconomics, macroeconomics, business cycle theory and optimal growth theory.[60]

Despite their clear advantages in the field of climate change research OLG models are still dominated by representative agent (RA) models. This has historical and computational reasons. On the one hand, RA models have a long tradition in economic growth theory and on the other hand, they come at considerably lower computation costs as they can describe competitive equilibria with streamlined methods of dynamic optimization.[61] Furthermore, proponents of the Ramsey model argue that this is the *"natural approach for studying the allocation of assets and resources across generations"* and that equity issues are adequately taken into account by comparing the sums of each generation's discounted utility levels for different consumption paths.[62]

Though OLG and RA differ clearly in their conceptual framework they must not necessarily produce different outcomes. For example, as Howarth (1998, 2000) and Stephan et al. (1997) show, infinitely-lived agent models and overlapping generations models can be calibrated to yield the same outcomes for climate policy under certain assumptions. They either postulate a certain degree of altruism of individuals or explicitly distinguish between private and social time preferences. Barro (1974) shows that for appropriate assumptions about altruism, finitely lived overlapping generations aggregate into a RA model. The assumption that social and private preferences diverge receives empirical support from Pope and Perry (1989), Luckert and Adamowicz (1993) and Lazaro et al. (2001). They provide empirical evidence that individuals apply lower rates of time preference in social contexts compared to private ones. In contrast, it is doubtful whether assumptions regarding the bequest motive of individuals stand the test of reality as empirical studies from Hurd (1987, 1989), Kopczuk and Lupton (2007) and Laitner and Ohlsson (2001) indicate that such motives are rather weak or insignificant.

Even if, under certain circumstances OLG and RA models yield the same outcomes, there are good reasons to explicitly consider the life cycles of different generations. Schneider et al. (2012), for instance, indicate several shortcomings of the RA approach: First, the preference parameters of the pure time preference rate of an RA planner is higher than the rate of the households in an equivalent OLG economy. Second, the RA

---

[60] See Tvede (2010), pp. 1–2.
[61] See Howarth (1998), p. 136.
[62] See Gerlagh and van der Zwaan (2001), p. 2.

model lacks explanatory power with respect to the question of a trade-off between inter- and intragenerational equity, as it does not consider finite life spans and the existence of overlapping generations. Last, because the RA economy assumes the social planner to be unconstrained, they are incapable of capturing second-best aspects of optimal policies.[63]

All in all, it is plausible to assume that neither the *consumer sovereignty* approach nor the *intergenerational equity* approach produce satisfying outcomes. Instead, a discounting scheme that takes into account the demographic structure of society (as it is the case in the framework of OLG models) might be more likely to reflect society's preferences accurately. In the next section, we want to derive such a discounting scheme and present its implications for optimal climate policy.

---

[63] See Schneider et al. (2012), p. 1635.

5. Intergenerational Discounting in the DICE Model

5.1. Derivation of the Discount Rate

In the following, we introduce a discounting scheme that is consistent with *consumer sovereignty* and *intergenerational equity*. This scheme stems from Krysiak (2010) and is derived in an OLG framework, assuming that individual's life-spans extend over a certain number of periods.[64] To be as close as possible to the RA framework, Krysiak assumes that all individuals have additive time-separable preferences with a constant discount rate. Individual utility is a function of consumption such that utility in period t is $U(c_t)$ where c_t is the individual consumption in period t.[65] The expected welfare for an individual born in period i with the fixed and finite time horizon T can be expressed by

Equation (1)

$$w_i := \sum_{t=i}^{T} \frac{\phi_{i,t-i}}{(1+\rho)^{t-i}} U(c_t)$$

where $\phi_{i,x}$ denotes the probability that an individual born in i still lives in period $i + x$. In period i, g_i individuals are born. To distinguish between different individuals born in the same period, we introduce the index j. That means that $w_{i,j}$ expresses the expected lifetime welfare of the jth individual born in period i. The consumption stream of individual (i,j) is denoted by $_{i\ldots T}c_{i,j}$. Last, the maximal possible lifetime of an individual born in period i is given by $\hat{T}_i$, such that $\phi_{i,x} = 0$ for all $x > \hat{T}_i$.

In analogy to Stern (2007), it is assumed that there is some probability θ that mankind does not procreate. Thus, $(1 - \theta)^i$ denotes the probability of continuing existence of a generation born in i. The realization of whether generation $i + 1$ exists or not is given by θ_i. With regards to this uncertainty, Krysiak denotes a realization of events by S. It describes the successive realizations of θ_i beginning in $i = 0$ till either procreation comes to an end or the time horizon T is reached.

Krysiak derives the discounting scheme based on several normative assumptions which become necessary to (1) guarantee for both *consumer sovereignty* and *intergenerational*

---

[64] Note that our derivation is based closely on that of Krysiak (2010).
[65] Krysiak defines utility as a function of consumption in period i, that is $U(c_i)$ and denotes the period in which an individual is born with t. Later on, this notation is reversed. For consistency reasons, we stick to the second notation.

equity, (2) to simplify the set up and (3) to be as close as possible to the RA approach. The following normative assumptions are applied:

(1) **Consumer Sovereignty (CS):** Social welfare is a function of the expected lifetime welfare of the individuals, which is measured according to their preferences.

Equation (2)

$$W\left({}_{0\ldots T}c_{0,1},\ {}_{0\ldots T}c_{0,2},\ \ldots,\ {}_{T\ldots T}c_{T,g_T}\right)$$

$$= W\left(w_{0,1}\left({}_{0\ldots T}c_{0,1}\right),\ w_{0,2}\left({}_{0\ldots T}c_{0,2}\right),\ldots,\ w_{T,g_T}\left({}_{T\ldots T}c_{T,g_T}\right),\right.$$

This assumption implies that individuals evaluate consumption streams only according to their preferences and that the social welfare function is only based on the individuals' intertemporal preferences (the relative evaluation of consumption in different points). It allows aggregating the evaluations of different individuals by a social welfare function (SWF) and respects individual's evaluation of consumption at the same time.

(2) **Stochastic Weak Anonymity (SWA):** For each realization S: If individuals (i, j) and (k, l) exist in this realization, the social welfare assigned to S is indifferent with respect to a permutation of these individuals' expected lifetime welfare. SWA guarantees that all individuals (present and future), once they exist, are given the same weight in the optimization problem.

(3) **Expected Welfare Maximization (EWM):** Social welfare equals the expected value of the social evaluations of all possible realizations S. Consequently, the weight given to the welfare of a future individual born in t should be the same as the probability that an individual is born.

(4) **Utilitarianism (U):** For each realization S, we assume that the social evaluation of S is utilitarian. That means that the realization S can be expressed by a weighted sum of the expected lifetime welfare of the individuals that exist in this realization. This assumption simplifies further analyses.

(5) **Identical Individuals (II):** All individuals exhibit the same individual time preference ρ and have the same utility function $U(c)$. Furthermore, it is assumed that all individuals that live in period t receive the same level of consumption c_t.

(6)   **Constant Population (CP):** It is assumed that in all periods the same number g of individuals is born and the probability $\phi_{i,x}$ is constant over time, that is, $\phi_{i+z,x} = \phi_{i,x} =: \phi_x$ for all $i, x, z \geq 0$. By CP, demographic change is excluded.

(7)   **Simple Life Duration (SLD):** For all $i \in 0, \dots, T$ there is some $\hat{T}_i > 0$, so that

Equation (3)

$$\phi_{i,x} = \begin{cases} 1 & 0 \leq x \leq \hat{T}_i \\ 0 & otherwise. \end{cases}$$

Individuals born live with certainty during their individual life duration $\hat{T}_i$. This assumption reduces the process to the change in the number of births g_i and the livespan $\hat{T}$.

By CS, the social welfare function can be expressed by

Equation (4)

$$W = W(w_{0,1}, \dots, w_{0,P_0}, w_{1,1}, \dots, w_{1,g_1}, w_{2,1}, \dots, w_{T,g_T})$$

where $w_{i,j}$ gives the individual welfare of the jth born individual in period i. Whereas g_i refers to the individuals born in period i, P_0 refers to the number of individuals alive in period 0.

Considering a fixed realization S of a succession of generations and taking into account assumption U, the social welfare can be written as

Equation (5)

$$W^S = \sum_{j=1}^{P_0} \alpha_{0,j}^S w_{0,j} + \sum_{i=1}^{T^S} \sum_{j=1}^{g_i} \alpha_{i,j}^S w_{i,j}.$$

$\alpha_{x,y}^S$ are some weights and T^S is the last existing generation in the realization S or the time horizon, whichever value is smaller. By assumption SWA, as all individuals who are considered in the realization S are known to exist, they should enter with the same weight, which allows us to normalize the weights to one. Together with EWM - the social welfare W is the expected value of W^S- we can deduce:

Equation (6)

$$W = \sum_{i=-\hat{T}_0}^{T} \sum_{j=1}^{g_i} (1-\theta)^{Max\{i,0\}} w_{i,j}.$$

The outer sum aggregates all periods from period $-\hat{T}_0$ to T. By setting $-\hat{T}_0$ as the lower limit of the outer sum, we consider everybody in our welfare optimization problem, even those who were born before $t=0$ but still live in $t=0$. The upper limit of the outer sum is given by T, which is the total time horizon. The inner sum adds up all individuals born in one period. In period i, g_i individuals are born. $(1-\theta)$ is the counter probability of no further procreation which is the probability that a generation is born in period i. The likelihood of extinction is increasing from period to period which is represented by the exponent $Max\{i,0\}$. For individuals born before or in our initial period their existence is definite. Thus, here the exponent becomes zero and $(1-\theta)$ becomes one. All other generations are discounted with the probability of existence. $w_{i,j}$ is the expected lifetime welfare of the j-th individual born in period i.

Under II, CP, SLD and by substituting Equation (1), (6) can be simplified to:

Equation (7)

$$W = \sum_{i=-\hat{T}}^{T} \sum_{t=Max\{i,0\}}^{Min\{\hat{T}+i,T\}} \frac{g}{(1+\rho)^{t-Max\{i,0\}}} (1-\theta)^{Max\{i,0\}} U(c_t).$$

From the Simple Life Duration (SLD) and the Constant Population (CP) assumptions it follows that the total sum of individuals born in one period can be represented by g. Identical Individuals (II) implies that all individuals have the same discount rate ρ, the same utility function $U(c_t)$ and reach the same level of consumption c_t in period t. Now the outer sum aggregates all individuals. Under the assumption that all people born before $t=0$ have the same life-span as the people born in period $i=0$ we can replace $-\hat{T}_0$ by $-\hat{T}$.[66] The inner sum is defined over the integration limits of $Max\{i,0\}$ and $Min\{\hat{T}+i,T\}$. It sums up the total utility streams of an individual born in i.

---

[66] It is important to clarify the difference between i and t. Whereas the former corresponds to the point of time at which some individual is born, the latter corresponds to the point of time at which consumption takes place. Thus, the outer sum starts at $i=-\hat{T}$ and the inner sum $t=Max\{i,0\}$. The upper limit of the inner sum ensures that people's consumption stream does not continue beyond their life duration or the end of our time horizon ends, respectively.

Resorting and evaluating the outer sum gives us:[67]

Equation (8)

$$W = \sum_{t=0}^{T} \frac{1}{\hat{T}+1} \left(\frac{g}{(1+\rho)^t} Max\{\hat{T} - t + 1, 0\} \right.$$

$$\left. + g(1-\theta)^t \frac{(1-\theta)(1+\rho)}{(1-\theta)\rho - \theta} \left(1 - \frac{1}{\left((1-\theta)(1+\rho)\right)^{Min\{t,\hat{T}+1\}}} \right) \right) U(c_t)$$

From which it follows that aggregated welfare in period t should be discounted as follows:

Equation (9)

$$d_{t=} \frac{1}{\hat{T}+1} \left(\frac{g}{(1+\rho)^t} Max\{\hat{T} - t + 1, 0\} \right.$$

$$\left. + g(1-\theta)^t \frac{(1-\theta)(1+\rho)}{(1-\theta)\rho - \theta} \left(1 - \frac{1}{\left((1-\theta)(1+\rho)\right)^{Min\{t,\hat{T}+1\}}} \right) \right)^{68}$$

The discounting scheme divides the discounting motive into two terms: The first term discounts all individuals that live in the initial period 0 which includes all individuals born before and in $t = 0$. As these people live with certainty, we only discount with their time preference. $Max\{\hat{T} - t + 1, 0\}$ weights the first term in comparison to the overall population existent in t. For example, in $t = 0$ it is one and in $t > \hat{T}$ - when the last generation from the starting period has passed away - it is zero. The second term inserts the discounting motive for all people that were not born in the starting period.

---

[67] For a detailed derivation see Appendix C.
[68] Note that the proposed discount rate is derived for a constant population, a simple and constant life-span and identical, individual time preferences ρ. In our analysis, we do not alter these assumptions and apply the discounting scheme directly in the DICE model. One should be aware that modifications in these assumptions can have substantial impact on the result of the discounting scheme. For example, Krysiak (2010) shows that the discounting scheme assigns considerably more weight to future generations once it is taken into account that the world population is likely to grow by about 50 to 70 percent during the next 50 years.

These do not live with certainty which is taken into account by $(1 - \theta)$, the probability of existence. Because the second term is the result of major simplifications, no further interpretation of it is possible.

What does this imply for the utility discounting? Figure 1 shows how the discounting scheme develops over the time. It turns out that the discount rate can be divided into three parts. In the first part, the initial group of generations strongly dominates the discounting motive. Because they live with certainty, we discount their utility only with the pure rate of time preference as suggested by the *consumer sovereignty* approach. Over the time the initial group receives increasingly less weight as the well-being of future generations becomes more and more important. During this second part utility is still discounted at a comparatively high rate, but not as heavily as suggested by the *consumer sovereignty* approach. Once the last person from the initial group passes away, the scheme enters the third part. Here, the discounting scheme follows an almost linear pattern similar to the *intergenerational equity* approach. In this phase, the curved progression of the discounting scheme is exclusively due to the risk of existence; the pure rate of time preference and the life-span do not influence the curvature of the discounting weight. This is quite a surprising result with great implications for optimal climate policy: From our sensitivity analysis (see section 5.3) we learn that savings behavior (and by that optimal climate policy) is only affected by relative differences in the discounting weights. Absolute differences do not matter. Because in the third part changes in ρ and $\hat{T}$ cause only absolute changes in the discounting weights we conclude that under the proposed discounting scheme time preferences do not influence long-term climate policy choice. In this regard, the proposed discounting scheme treats all future generations equally. The only reason to shift consumption from very far future generations towards not so far future generations is that the former come to existence with a smaller probability.

This has far-reaching implications for the normative concept of the proposed discounting scheme: First, today's and soon to-be-born generations are endowed with the same weights as under the *consumer sovereignty* approach. Over the time individual preferences receive increasingly less weight and in the long-run they do not matter at all. Then, only changes in the risk of existence motivate changes in climate policy paths, as proposed by the *intergenerational equity* approach. Apparently, the discount rate does not incorporate *consumer sovereignty* and *intergenerational equity* simultaneously, but rather applies them successively. Thus, it seems questionable whether the proposed discounting scheme really solves the *consumer sovereignty* and *intergenerational equity* dilemma. We do not generally reject its outcome. However, it must be clear that one still has to take an ethical stance no matter what the applied discount rate is.

5.2. Results

Next, we present the main implications for climate policy if one applies the proposed discounting scheme. Let us emphasize again that Integrated Assessment Models such as DICE-2013R do not produce "true" outcomes. They should rather be understood as tools that help to understand the complex relationships of different systems. Thus, our results should not be over-interpreted. Even if, they seem to be very precise, they are not and at best, they can only indicate different policy paths. For better understanding, we contrast our results with a number of other scenarios:

Baseline: The *Baseline* scenario describes what would be the outcome if current climate policies were extended indefinitely. Thus, it describes a world in which no further efforts to reduce the impacts of climate change are made apart from those that are already agreed upon. In the DICE-2013R model, it is assumed that policies were the equivalent of $1 per ton of CO_2 in 2010 and that the global carbon price will increase by 2 percent per year till 2200. The consideration of such a baseline scenario is common in economic forecasting and delivers a good benchmark for potential policies.

Nordhaus: This scenario maximizes economic welfare under the *consumer sovereignty* approach: the parameters are set in such manner that the discount rate of consumption reflects the interest rate on the market. Accordingly, the pure rate of time preference is set to 1.5 percent and the elasticity of the marginal utility of consumption to 1.45 Percent.

Stern: In contrast to *Nordhaus*, in the *intergenerational equity* approach utility discounting due to time preferences is rejected. The only reason for assigning a positive value (0.1 percent) to the pure rate of time preference is because of the risk of extinction. As proposed by the Stern Review, the elasticity of marginal utility of consumption is set to 1.

Krysiak: *Baseline*, *Stern* and *Nordhaus* serve as benchmark scenarios for the proposed scenario which we call *Krysiak*. *Krysiak* advocates the discounting scheme derived above, combining both *consumer sovereignty* and *intergenerational equity*. It comes with three parameters: the pure rate of time preference ρ, the risk of extinction θ and the expected life duration $\hat{T}$. To the pure rate of time preference we assign a value of 1.5

percent. This is an arbitrary choice. However, what speaks for it is that it delivers the highest comparability with the alternative scenarios. θ is set to 0.1 percent, again for reasons of comparison, as this is the value adopted by Stern. Nevertheless, it could be also argued that the risk of extinction should be higher. For example, the Future of Humanity Institute estimates the overall risk of human extinction before 2100 to be 19.0 percent.[69] The corresponding yearly probability is around 0.2 percent. The life duration is set to $\hat{T} = 75$ as we assume it to be represented best by the world's average life expectancy in the mid-century.[70] The elasticity of the marginal utility of consumption is set to 1.45.

Lim2t: Furthermore, we design a scenario which implements a temperature constraint. *Lim2t* applies the discounting scheme of *Krysiak*, our choice of parametrization plus the constraint that global temperature must not exceed 2°C above the 1900 average. Determining optimal climate policy for this constraint is particularly interesting since there is a broad scientific and political consensus that temperature changes above this threshold will involve unpredictable risks and uncertainties for the global economic and environmental system.[71] Furthermore, the major part of the world's political community has agreed on pursuing the objective of limiting climate change to 2°C.[72]

Note that in all scenarios we assume full participation of all nations and no climate constraints.[73] Based on this assumption, we obtain the most efficient policy outcomes for the underlying scenarios since they balance the present value of the abatement costs and the present value of the benefits of reduced climate damage. Although such assumptions are unrealistic, they become necessary to derive policy results that can serve as efficiency benchmarks.[74]

Table 1 gives a summary of the results of the different policy scenarios. The first column shows the present value of consumption in trillions of USD. Column 2 and 3 show the net economic benefit (total consumption) of the different policy scenarios, the former presenting the benefit in absolute terms and the latter the benefit in percentage points relative to the *Baseline* scenario. All policies result in net economic welfare gains. The largest gain ($60.9 trillion) in consumption is obtained from implementing the optimal

---

[69] See Sandberg and Bostrom (2008), p. 1.

[70] See UN (2012), p. 38.

[71] See IPCC (2014), p. 11.

[72] See Knopf et al. (2012), p. 121.

[73] "No climate constraints" means no limits for atmospheric CO_2 concentrations or global temperature increase are built into the scenarios.

[74] See Nordhaus and Sztorc (2013), p. 24.

policy according to *Krysiak* which corresponds to 2.42 percent of the present value of the *Baseline* scenario. Implementing *Nordhaus* and *Lim2t* climate policies generate welfare benefits of $56.7 trillion (*Nordhaus*) and $53.3 trillion (*Lim2t*), respectively. With $25.5 trillion *Stern's* policy provides for the smallest gain in total present value consumption.[75]

Table 1. Results of Major Runs for the DICE-2013R Model

Policy	Present value of consumption	Difference from *Baseline*	Difference from *Baseline*	Social Cost of Carbon	Carbon Price		Global Temperature Change	
				2010	2015	2100	2100	2150
	(Trillions of 2005 U.S. $)		(% of *Baseline*)	(2005 U.S. $ per Ton Carbon)			(°C from 1900)	
Baseline	2522,32	0,00	0,00%	15,5	1,1	5,9	3,85	5,36
Nordhaus	2581,00	58,68	2,33%	14,7	17,7	142,8	3,09	3,30
Stern	2547,81	25,48	1,01%	75,4	89,8	218,1	2,04	1,96
Krysiak	2583,26	60,93	2,42%	27,5	34,2	218,1	2,49	2,39
Lim2t	2575,60	53,27	2,11%	43,3	54,6	218,1	2,00	1,92

Column 4 shows the "social cost of carbon" in 2010 and Column 5 and 6 show the "carbon price" in 2015 and in 2100. Whereas the social cost of carbon measures the economic damage for society associated with emitting one further metric ton of carbon dioxide, the carbon price is the price that must be paid by polluters for one further ton of carbon. The carbon price is also called "carbon tax" because it can be implemented either by a trading regime or by a taxation policy. The price of carbon induces efficiency when it offsets the increasing cost of reducing carbon emissions with the increasing benefits of reducing climate damage. This state is reached when the carbon price equals the social cost of carbon.[76] The social cost of carbon under *Krysiak* is $27.5 per ton CO_2 emission in 2010, which lies between Stern ($75.4) and *Nordhaus* ($14.7). The optimal carbon price of $34.2 in 2015 for *Krysiak* also lies between *Stern* ($89.8) and *Nordhaus* ($14.7). In 2100, the optimal carbon price under the policies of *Stern* and *Krysiak* is the same ($ 218.1), whereas the price of *Nordhaus* is $142.8. The last column of Table 1 shows the projected global mean temperature change in °C in 2100 and 2200.

---

[75] Though such comparisons are commonly made in the field of climate change economics, the explanatory power is limited. For reasons of comparison the underlying discount factor must be identical in all scenarios. For example, here all scenarios are discounted with the consumption discount rate of *Krysiak*. As all other policies besides *Lim2t* are derived from different discount rates, such discounting must come at their disadvantage and present values are likely to be underestimated.
[76] See Nordhaus (2008), p. 81.

Implementing the policy advised by *Krysiak* would result in a temperature increase of 2.04 °C by 2100.

Carbon Prices and Emissions Controls

Table 2-1 and Figure 2-1 show the prices per metric ton of carbon dioxide associated with different policies. It is assumed that carbon prices are harmonized across regions, which could take place either through uniform taxes or through a trading system.[77] For the initial time period 2010 the carbon price is set to $1 per metric ton of carbon dioxide for all scenarios as this value corresponds to the CO_2 price of already existing policies. From Figure 2-1 we see that all scenarios (*Krysiak*, *Stern*, *Nordhaus* and *Lim2t*) call for higher carbon price paths than the *Baseline* scenario. In the *Stern* scenario, the optimal carbon tax rises immediately and reach a value of more than $100 by 2020. Until the mid of this century it remains above all other scenarios and reaches its maximum in 2070 ($253). Only the optimal carbon price path of *Lim2t* exceeds *Sterns* optimal carbon price: In comparison to *Stern*, in *Lim2t* the well-being of today's generations receives more weight in our considerations and thus consumption must shift from the future towards now. This translates into less climate protection in the early years but, as the two degree limit is binding, a steep and ongoing rise in carbon tax must follow a few years later to compensate for the first years of relatively low carbon prices. *Lim2t* exceeds the *Stern* path in 2045 ($178.5) and lies above *Stern* till 2075. The *Kryslak* scenario implies an optimal taxation path that always stays below or on the same level as *Stern* and *Lim2t* and peaks in 2085 at $232.7. In contrast, the optimal carbon tax under *Nordhaus* lies well below all other paths during the next 100 years, reaching its maximum at the end of the century (2100: 142.8).

Table 2-2 and Figure 2-2 present the emissions-control rates for the five scenarios. The emissions-control rate describes the extent to which industrial emissions are reduced below the level which would be reached without any reduction. The curve progression is very close to that of the optimal carbon price: The *Stern* scenario calls for the fastest climb in the rate path which is then overtaken by *Lim2t* in 2045. Both *Stern* and *Lim2t* call for a 100% emissions-control rate (zero industrial emissions) before 2075. In the *Krysiak* path the emissions-control rate stays below *Stern* and above *Nordhaus* until 2090. It starts at 32 percent in 2020 and rises to 61 percent in 2050. In 2090, it reaches

---

[77] See Nordhaus (2008), p. 81.

the 100 percent reduction mark. Until the end of this century the optimal rate under the *Nordhaus* discounting is always below the other scenarios.

Industrial Emissions, CO_2 Concentrations, and Global Warming

Next, we consider the implications of the five scenarios for industrial emissions, CO_2 concentrations and atmospheric temperature change. Table 2-3 and Figure 2-3 show the aggregated CO_2 emissions. Projections of the *Baseline* scenario show that without any further climate policy, the aggregated emissions would rise continuously, reaching 102.5 gigatons of CO_2 ($GtCO_2$) by the end of the century. All scenarios except *Baseline* follow a hump-shaped pattern after a small and immediate cut-back in emissions. Size and height of the hump depends on the scenario: Whereas under *Nordhaus* the hump reaches its climax not before mid-century (2055: 46.4 $GtCO_2$), exceeding the initial amount of 33.6 $GtCO_2$ in 2010 by more than 12 $GtCO_2$, the *Stern* scenario implies a cutback of 10 $GtCO_2$ already at its maximum of 22.9 $GtCO_2$ in 2025. *Krysiak, Lim2t* and *Stern* call for a 100 percent cut-back of total carbon dioxide emissions before 2100.

The atmospheric concentrations of CO_2 are shown in Table 2-4 and Figure 2-4. The initial concentration in 2010 was 390 parts per million (ppm). In the *Baseline* scenario, they rise permanently till the mid of the next century, reaching a value of 1134 ppm in 2150. The CO_2 concentration paths of *Lim2t* and *Stern* develop in a similar fashion. Both increase only slightly till the mid of this century. Having reached their maximum in 2050 (*Stern* at 447 ppm, *Lim2t* at 453 ppm), they decline again. In 2150, their concentration level is below their initial levels of 2010 (*Stern* at 388 ppm, *Lim2t* at 384). In the *Nordhaus* scenario, the atmospheric CO_2 concentrations are above all other scenarios and continuously increase until they reach 610 ppm by 2110. Again, the *Krysiak* CO2 concentration curve follows a path just between *Nordhaus* and *Stern*.

Table 2-5 and Figure 2-5 show the increase in global mean temperature in °C. It turns out that even with strong and immediate climate policies, as being proposed by *Stern* and *Lim2t*, a major temperature increase in the next 40 years cannot be prevented. This phenomenon is due to the emissions from the past because the climate system's reaction to decreasing output of carbon dioxide emissions is time-lagged and slow.[78] For comparison, the difference in temperature change between the *Baseline* scenario and the strongest policy (*Lim2t*) is only 0.31°C (2.01°C-1.70°C) in 2050. However, after the

---

[78] See Nordhaus (2008), p. 105.

mid of this century the implications of today's actions become apparent: Without any further cut in CO_2 emissions the global mean temperature will rise to almost 4°C by the end of the century (2100: 3.85°C). In contrast, for *Lim2t* and *Stern* the temperature change will not or only slightly exceed the 2°C limit. In the *Krysiak* scenario, the higher increase in mean temperature is also comparatively small, never rising above 2.5°C until 2150.

Consumption and Economic Output

The main difference between the discounting schemes of *Stern* and the others is that in the former, time preferences are not taken into account. Stern discounts utility only because there is uncertainty about the existence of future generations. Here, in comparison to the other discounting schemes, future generations receive more weight relative to today's which calls for a consumption shift towards the future. Table 2-6 and Figure 2-6 present the per capita consumption for the different scenarios. In the first decades, *Stern*'s optimal per capita consumption path is distinctly lower than the others, starting with an initial difference of more than $700 in 2010. Till mid-century it gradually catches up with all other scenarios. Along with *Lim2t*, it allows for the highest per capita consumption paths from 2075 till the end of 2150. The initial consumption gap is due to the comparatively high savings rates and abatement costs. Because *Stern*'s discounting prioritizes the well-being of future generations measures with long-term benefit for economic development like capital investment and climate protection will become relatively more attractive.

In comparison to *Nordhaus*, *Lim2t* and *Krysiak* also call for a bigger shift in consumption towards future generations. In the long-run their per capita consumption paths develop similarly to the *Stern* scenario. Because *Lim2t* implements the temperature constraint, it calls for distinctly higher abatement efforts than *Krysiak*. A high volume of abatement costs expenditure provides for sustaining the temperature limit at the cost of slight consumption losses. The *Baseline* path develops contrary to *Stern*, *Lim2t* and *Krysiak*. Initially, it starts from the highest level (2010: $6,886) but it is then overtaken by all other scenarios by the end of this century. In 2150, the consumption gap between *Lim2t* and the *Baseline* is marked: Whereas in the former people enjoy a per capita consumption of $70,185 on average, in the latter it would be only $64,123. As predicted in the *Nordhaus* scenario the consumption path starts on a high level (2010: $6,878), quite similar to *Baseline*, but decreases relative to the other scenarios over time. In 2150, per capita consumption would reach a value of $67,609.

Projections for the development of global output are similar to the projections of consumption paths. Table 2-7 and Figure 2-7 show the net gross domestic product in trillions which is the global output after abatement and damage costs. In the *Baseline* run, net output develops with the slowest growth rate, reaching a value of $491 trillion in 2100. In contrast, with *Stern* the GDP reaches its largest development due to high savings rates. By the end of this century, it is about $1050 trillion. Though *Nordhaus'* output path is distinctly higher than in the *Baseline* scenario, it is still the lowest compared to *Stern*, *Lim2t* and *Krysiak*. *Krysiak* and *Lim2t* show almost equal growth paths in net outcome.

5.3. Sensitivity Analysis

To test our results for robustness, we conduct a sensitivity analysis. This becomes necessary because there is uncertainty regarding the "true" value for the discounting scheme. In the reference scenario (*Krysiak*), we assigned a value of 1.5 percent for the pure rate of time preference, 0.1 percent for the risk of existence and 75 years for the expected life-span. These, however, were only best guesses and to some extent arbitrary choices. To analyze to which extent the models' outcome will react to different, also plausible choices of these parameters we conduct an OFAT sensitivity analysis. This approach is the simplest version of sensitivity analysis and explores the effect on the output if one changes one-factor-at-a-time (OFAT), keeping all other factors constant. In the field of climate modeling this approach is very common as it minimizes computational costs and allows for identifying the sole effect of the factor change on the output.[79] It should be mentioned, however, that this approach also has considerable drawbacks that have been recently pointed out by various researchers. Its main disadvantage is that it cannot detect the interactions between the different input variables.[80] Nevertheless, we claim that the OFAT method is sufficient for our purposes as it is standard in integrated-assessment modeling and other methods would go beyond the scope of this work.

---

[79] See Saltelli and Annoni (2010), p. 1510.
[80] See Anderson et al. (2014), p. 272.

Before we turn to our results, a few words should be said about the factor ranges that are applied in this analysis. This sensitivity analysis considers the percentage change in GDP losses[81] due to changes in either θ, ρ or $\hat{T}$, ceteris paribus. For the risk of extinction, we test for θ taking on values within the interval of 0.01 percent and 0.5 percent which corresponds to a minimum risk of one percent and a maximum risk of 35 percent that mankind will not survive until 2100. Forming exact estimates for this probability is extremely difficult as we do not know how likely certain life-threatening events are.[82] Rees (2003), Hepburn (2006) and Stern (2007) argue quite plausibly that this probability is almost certainly below 0.5 percent per year, which prompts us to choose the same upper limit. Setting $\theta = 0.001$ as the lower limit is again to some extent arbitrary. However, from our perspective, assuming a one percent probability for human extinction should be a good guess for the lower risk as it is far smaller than most studies estimate. For the interval of the pure rate of time preference ρ, we set a lower bound of $\rho = 1$ percent and an upper bound of $\rho = 4$ percent. This allows for a broad set of possible values that are distinctly higher and lower than the generally assumed range of 2 to 3 percent a year.[83] With respect to the expected live duration $\hat{T}$, we test for an interval with a minimum value of 69 years and a maximum value of 81 years. The former value corresponds to the average life expectancy of a person born today, the latter to the average life expectancy of someone born at the end of the 21st century.[84] Since $\hat{T}$ is the life expectancy of an average person born in our total time horizon, it seems reasonable to choose $\hat{T}$ within this interval. To keep the analysis simple, we are not applying any specific probability distribution for the input variables. Instead, we choose random values within the factors' interval and test the implications for the outcome.

Figures 3-1, 3-2, 3-3 present the results of our sensitivity analysis. First, we see that with rising values of θ net GDP gains from optimal policy relative to the *Baseline* policy decrease. If the probability of extinction increases, this means that future generations receive smaller weights in our welfare maximization. In consequence, future consumption and production becomes less important, investments in climate protection and capital accumulation is getting less attractive and for increasing values of θ the economic output decreases.[85]

---

[81] The percentage change in GDP loss: (*Baseline* GDP – *Krysiak* GDP) / *Krysiak* GDP.
[82] See Bostrom (2013), p. 15.
[83] See Dasgupta (2008), p. 157.
[84] See UN (2012), p. 38.
[85] Note that for different values of θ savings rates s diverge over time. The higher the risk of extinction the lower s (see Figure 4-1).

Figure 3-2 shows the per-capita GDP loss for different values of ρ. In analogy to θ, with increasing values of the pure rate of time preference GDP losses become smaller. It turns out that this implication will only be true for the first decades of our total time horizon. Moreover, once the rate exceeds the threshold of 1.5 percent[86] the GDP loss from *Baseline* policy reverses to a net gain in the first decades. At first glance, this result might seem a bit odd; however, it is easily explained by the structure of the underlying discounting scheme. As shown in section 5.1, the discounting scheme proposed by Krysiak (2010) treats people differently with respect to the time of birth. People who live with certainty build up the first part of the scheme. They are only discounted with the pure rate of time preference. The second part contains all future-born generations whose existence is uncertain. In the first decades, let us say until 2050, the first group of people dominates the discounting motive simply due to its relative number of people. With increasing values of ρ, their motive to shift consumption towards the present dominates more and more the converse motive of future generations. Consequently, with rising values of ρ, savings rates are lowered to allow higher immediate consumption levels. Figure 3-5 shows the development of the savings rates for different choices of ρ and supports our theory. The higher ρ, the lower the savings rate s. From 2050, the preferences of the "certain" generations fade out and consequently the discounting factors develop equally for different values of the pure rate of social time preference in relative terms (see Figure 4-2).[87] Therefore, savings rates converge to the same path and in the long-run per capita GDP must also converge to the same level, implying identical curve progression for the percentage loss in GDP per capita.[88]

Similar to ρ, an increase in the life expectancy implies a decrease in GDP loss but only for the period from 2010 until 2090 (see Figure 3-3). Equation 9 shows that $\hat{T}$ plays a significant role only in the first part of the discounting scheme. The older today's people become, the more weight their preferences are given to in our welfare considerations. Again, because present generations prefer to shift consumption towards their life period,

---

[86] The threshold is given by $\rho = 1.5$ because in the *Baseline* scenario the pure rate of time preference is set to the same value. Any value above 1.5 implies higher savings rates than in the *Baseline*, and vice versa.

[87] Once the first generation passes away, all future generations are treated relatively equally for different values of ρ. An increase in ρ only influences the level of the discounting scheme, but not its change. As all people are treated the same, their preferences are treated similarly in relative terms. Consequently, savings rates must be identical, too.

[88] If the savings rates are identical, capital accumulation must increase in those growth paths that initially had smaller savings rates. Because technological change and population development is exogenous in this growth model and capital is the only input variable besides the climate system, all scenarios must reach the same level of output for identical capital stocks.

increasing values in $\hat{T}$ imply smaller savings rates for the first time period. After that time, the well-being of the initial generations no longer matters and savings rates converge to the same value in all cases (see Figure 3-6). Overall, changes in the expected life duration have only very small impact on short-term GDP development.

From our sensitivity analysis, we conclude that our results are robust to exogenous changes of parameters. Changes in the parameter values cause logical and expected changes in our output variables. In the long-run, the risk of extinction is the only variable that has a sustainable impact on our policy outcomes. With a maximum difference of 2.6 percentage points between the scenario assuming the highest and the lowest risk, this impact is fairly small.

6. Conclusions

Both *consumer sovereignty* and *intergenerational equity* are eligible stances in the discussion on how to weigh future generations against present ones. At best, they should not be played off against each other. Under the representative agent approach, however, one has to take sides with major implications for climate policy. Krysiak (2010) shows that this problem can be overcome in the framework of an overlapping generations model. For a set of normative assumptions, he derives a utility discounting scheme which assigns to all generations equal weights without disregarding their true preferences. In this thesis, we applied this discounting scheme in the latest DICE model and presented its implications for optimal climate policy.

As expected, the *intergenerational discounting* approach (*Krysiak*) calls for a climate policy path which is between the ones of *consumer sovereignty (Nordhaus)* and *intergenerational equity (Stern)*. Under *Krysiak*, in 2015, the carbon price should be set to $34.2 per metric ton of carbon dioxide and continuously increased to $237.7 by 2085. Afterwards, it should be slowly reduced so that it reaches a price of $218.1 in 2100. Under *Stern* the carbon price should be set to a distinctly higher initial value (2015: $89.9) and increase strongly. Similarly, under *Krysiak*, after peaking in 2070 it declines again, reaching a value of $218.1 by the end of this century. Under *Nordhaus*, in contrast, the initial carbon price is to be set to a comparatively lower value of $17.7 in 2015. It then increases, following a path which stays below the two other scenarios. In 2100, it reaches its maximum carbon price $142.8, which is about $75 below *Stern* and *Krysiak*. Under *Krysiak*, the optimal growth path implies a carbon concentration level of 467 ppm by the end of this century. The atmospheric temperature increases to 2.49 °C above preindustrial level, which is about 0.5 degrees above *Stern* (2.04 °C) and 0.6 degrees below *Nordhaus* (3.09 °C).

In section 5.3, we analyzed how variations in the parameters θ, ρ and $\hat{T}$ influence the policy outcome. It turned out that the pure rate of time preference and the expected life-span have only a short-term impact on savings rates and economic growth, whereas the risk of extinction also affects long-term growth. We concluded that the discounting scheme proposed by Krysiak does not incorporate both concepts simultaneously but rather applies them successively. In this regard, the discounting scheme seems to fail in solving the *consumer sovereignty* and *intergenerational equity* dilemma in a satisfactory way. Further efforts to find a solution are needed and the framework of Krysiak (2010) should give conceptual guidance to such a solution.

Apart from that, we want to stress that policy outcomes based on integrated assessment modeling must be generally regarded with precaution. Recent literature indicate several key areas in which present IAMs suffer from crucial flaws in modeling and omit key factors such as potential tipping points, endogenous drivers of economic growth or realistic damage functions. As long as these flaws exist and key factors are omitted, climate policy derived from analysis as performed in this thesis is likely to be misleading and incorrect.

This does not imply that policy makers should abstain from further actions to reduce carbon emissions and wait until further research has been undertaken. In our opinion, just because we cannot accurately estimate the consequences of our actions does not mean that we are not responsible for them. Instead, climate policy should follow a precautionary approach which ensures that certain unlikely but devastating events can be almost certainly excluded. IAMs scenarios such as the *Lim2t* which incorporate reasonable temperature constraints can give guidance for such policies.

References

Ackerman, F., Stanton, E. A. and R. Bueno (2010): *Fat Tails, Exponents, Extreme Uncertainty: Simulating Catastrophe in DICE*, Ecological Economics 69(8), 1657–1665.

Alley, R. B., Marotzke, J., Nordhaus, W. D., Overpeck, J. T., Peteet, D. M., Pielke, R. A., Pierrehumbert, R. T., Rhines, P. B., Stocker, T. F. and L. D. Talley (2003): *Abrupt Climate Change*, Science 299(5615), 2005–2010.

Anderson, B., Borgonovo, E., Galeotti, M. and R. Roson (2014): *Uncertainty in Climate Change Modeling: Can Global Sensitivity Analysis Be of Help?*, Risk Analysis 34(2), 271–293.

Annan, J. D. (2001): *Modelling Under Uncertainty: Monte Carlo Methods for Temporally Varying Parameters*, Ecological Modelling 136(2), 297–302.

Arrow, K. J., Cline, W. R., Maler, K.-G., Munasinghe, M., Squitieri, R. and J. E. Stiglitz (1996): *Intertemporal Equity, Discounting, and Economic Efficiency*, in Bruce, J. P., Lee, H. and E. F. Haites (eds.): *Climate Change 1995. Economic and Social Dimensions of Climate Change, Contribution of Working Group III to the Second Assessment Report of the Intergovernmental Panel on Climate Change*, Cambridge University Press, Cambridge, UK, 125–144.

Arrow, K. J., Cropper, M., Gollier, C., Groom, B., Heal, G. M., Newell, R. G., Nordhaus, W. D., Pindyck, R. S., Pizer, W. A. and P. Portney (2012): *How Should Benefits and Costs Be Discounted in an Intergenerational Context? The Views of an Expert Panel*, Technical Report, Resource for the Future Discussion Paper, Washington DC.

Barro, R. J. (1974): *Are Government Bonds Net Wealth?*, Journal of Political Economy 82(6), 1095–1117.

Bayer, S. (2000): *Intergenerationelle Diskontierung am Beispiel des Klimaschutzes*, Metropolis-Verlag, Marburg.

Bayer, S. (2003): *Generation Adjusted Discounting in Long-Term Decision-Making*, International Journal of Sustainable Development 6(1), 133–149.

Bayer, S. (2004): *Nachhaltigkeitskonforme Diskontierung - Das Konzept des "Generation Adjusted Discounting"*, Vierteljahrshefte zur Wirtschaftsforschung 73(1), 142–157.

Bostrom, N. (2013): *Existential Risk Prevention as Global Priority*, Global Policy 4(1), 15–31.

Buchholz, W. and J. Schumacher (2009): *Die Wahl der Diskontrate bei der Bewertung von Kosten und Nutzen der Klimapolitik*, in, Beckenbach, F., Leipert, C., Meran, G., Minsch, J., Nutzinger, H. G., Weimann, J. and U. Witt (eds.): *Diskurs Klimapolitik*, Jahrbuch Ökologische Ökonomik 6, Metropolis Verlag, Marburg, 1-33.

Cline, W. R. (1992): *The Economics of Global Warming*, Institute for International Economics, Washington DC.

Dasgupta, P. (2007): *Commentary: The Stern Review's Economics of Climate Change*, National Institute Economic Review 199(1), 4–7.

Dasgupta, P. (2008): *Discounting Climate Change*, Journal of Risk and Uncertainty 37(2-3), 141–169.

Gerlagh, R. and B. C. C. van der Zwaan (2001): *Overlapping Generations versus Infinitely-Lived Agent: The Case of Global Warming*, in Howarth, R. and D. Hall (eds.): *The Long-Term Economics of Climate Change 3*, JAI Press, Stanford, Connecticut, 301-327.

Gollier, C. (2006): *An Evaluation of Stern's Report on the Economics of Climate Change*, Institut d'Économie Industrielle, Working Paper No. 464, Toulouse.

Gollier, C. (2010): *Ecological Discounting*, Journal of Economic Theory 145(2), 812–829.

Hall, R. E. (1988): *Intertemporal Substitution in Consumption*, Journal of Political Economy 96(2), 339–357.

Harrod, R. F. (1948): *Towards a Dynamic Economics*, McMillan, London.

Henderson, N. and I. Bateman (1993): *Intergenerational Discounting: Public Choice and Empirical Evidence for Hyperbolic Discount Rates*, CSERGE, Working Paper No. GEC 93-02, London.

Hepburn, C. (2006): *Discounting Climate Change Damages: Working Note for the Stern Review*, Environmental Change Institute and Department of Economics, Working Paper, Oxford.

Howarth, R. B. (1998): *An Overlapping Generations Model of Climate-Economy Interactions*, The Scandinavian Journal of Economics 100(3), 575–591.

Howarth, R. B. (2000): *Climate Change and the Representative Agent*, Environmental and Resource Economics 15(2), 135–148.

Hurd, M. D. (1987): *Savings of the Elderly and Desired Bequests*, The American Economic Review 77(3), 298–312.

Hurd, M. D. (1989): *Mortality Risk and Bequests*, Econometrica 57(4), 779–813.

IPCC (2014): *Summary for Policymakers*, in Field, C. B., Barros, V. R., Dokken, D. J., Mach, K. J., Mastrandrea, M .D., Bilir, T. E., Chatterjee, M., Ebi, K. L., Estrada, Y. O., Genova, R. C., Girma, B., Kissel, E. S., Levy, A. N., MacCracken, S., Mastrandrea,

P. R. and L. L. White (eds.): *Climate Change 2014: Impacts, Adaptation, and Vulnerability. Part A: Global and Sectoral Aspects. Contribution of Working Group II to the Fifth Assessment Report of the Intergovernmental Panel on Climate Change.* Cambridge University Press, Cambridge, UK, 1–32.

van der Sluijs, J. P. (2002): *Integrated Assessment,* in Tolba, M. K. (ed.): *Encyclopedia of Climate Change,* John Willy & Sons, Chichester, 250–253.

Kelly, D. L. and C. D. Kolstad (1999): *Integrated Assessment Models for Climate Change Control,* in Folmer, H. and T. Tietenberg (eds.): *The International Yearbook of Environmental and Resource Economics 1999/2000: A Survey of Current Issues,* Edward Elgar Publishing, Cheltenham, 171–197.

Knopf, B., Kowarsch, M., Flachsland, C. and O. Edenhofer (2012): *The 2 C Target Reconsidered,* in Edenhofer, O., Wallacher, J., Lotze-Campen, H., Reder, M., Knopf, B. and J. Muller (eds.): *Climate Change, Justice and Sustainability: Linking Climate and Development Policy,* Springer Science + Business Media, Dordrecht, 121–137.

Kopczuk, W. and J. P. Lupton (2007): *To leave or not to leave: The distribution of bequest motives,* The Review of Economic Studies 74(1), 207–235.

Krysiak, F. C. (2010): *Discounting, Intergenerational Equity, and Demographic Change,* Department of Business and Economics, Working Paper, Basel.

Laitner, J. and H. Ohlsson (2001): *Bequest Motives: A Comparison of Sweden and the United States,* Journal of Public Economics 79(1), 205–236.

Lazaro, A., Barberan, R. and E. Rubio (2001): *Private and Social Time Preferences for Health and Money: An Empirical Estimation,* Health Economics 10(4), 351–356.

Luckert, M. K. and W. L. Adamowicz (1993): *Empirical Measures of Factors Affecting Social Rates of Discount,* Environmental and Resource Economics 3(1), 1–21.

Manne, A., Mendelsohn, R. and R. Richels (1995): *MERGE: A Model for Evaluating Regional and Global Effects of GHG Reduction Policies*, Energy Policy 23(1), 17–34.

Nordhaus, W. and P. Sztorc (2013): *DICE 2013R: Introduction and User's Manual.* Retrieved from http://www.econ.yale.edu/~nordhaus/homepage/documents/DICE_Manual_103113r2.pdf Accessed on March 02, 2015.

Nordhaus, W. D. (1997): *Discounting in Economics and Climate Change*, Climatic Change 37(2), 315–328.

Nordhaus, W. D. (2006): *The "Stern Review" on the Economics of Climate Change*, National Bureau of Economic Research, Working Paper No. 12741, Cambridge, Massachusetts.

Nordhaus, W. D. (2007): *A Review of the" Stern Review on the Economics of Climate Change"*, Journal of Economic Literature 45(3), 686–702.

Nordhaus, W. D. (2008): *A Question of Balance: Weighing the Options on Global Warming Policies*, Yale University Press, New Haven, Connecticut.

Paqué, K.-H. (2008): *Zins, Zeit und Zukunft - Zu Ökonomie und Ethik globaler Klimamodelle,* in Gischer, H., Reichling, P., Spengler, T. and A. Wenig (eds.): *Transformation in der Ökonomie. Festschrift für Gerhard Schödiauer zum 65. Geburtstag*, Gabler Verlag, Wiesbaden, 269–286.

Parson, E. A. and K. Fisher-Vanden (1997): *Integrated Assessment Models of Global Climate Change*, Annual Review of Energy and the Environment 22(1), 589–628.

Pigou, A. C. (1920): *The Economics of Welfare*, McMillan, London.

Pope, C. A. and G. Perry (1989): *Individual versus Social Discount Rates in Allocating Depletable Natural Resources over Time*, Economics Letters 29(3), 257–264.

Ramsey, F. P. (1928): *A Mathematical Theory of Saving*, The Economic Journal 38(152), 543–559.

Rawls, J. (1972): *A Theory of Justice*, Oxford University Press, Oxford.

Rees, M. J. (2003): *Our Final Century*, Heinemann, London.

Roemer, J. E. (2011): *The Ethics of Intertemporal Distribution in a Warming Planet*, Environmental and Resource Economics 48(3), 363–390.

Samuelson, P. A. and W. D. Nordhaus (1989): *Economics*, 13th, McGraw-Hill, New York.

Samuelson, P. A. (1958): *An Exact Consumption-Loan Model of Interest With or Without the Social Contrivance of Money*, Journal of Political Economy 66(6), 467–482.

Saltelli, A. and P. Annoni (2010): *How to Avoid a Perfunctory Sensitivity Analysis*, Environmental Modelling and Software 25(12), 1508–1517.

Sandberg, A. and N. Bostrom (2008): *Global Catastrophic Risks Survey*, Future of Humanity Institute, Oxford University, Oxford.

Schelling, T. C (1999): *Intergenerational Discounting*, in Portney, P. R. and J. P. Weyant (eds.): *Discounting and Intergenerational Equity*, Resources for the Future, Washington DC., 99–101.

Schneider, M. T., Traeger, C. P. and R. Winkler (2008): *Intergenerational Equity in Long-Run Decision Problems Reconsidered*, EARE, Gothenburg.

Schneider, M. T., Traeger, C. P. and R. Winkler (2012): *Trading Off Generations: Equity, Discounting, and Climate Change*, European Economic Review 56(8), 1621–1644.

Solow, R. M. (1970): *Growth Theory: An Exposition,* Oxford University Press, Oxford.

Solow, R. M. (1974): *The Economics of Resources or the Resources of Economics,* American Economic Review 64(2), 1–14.

Stanton, E. A., Ackerman, F. and S. Kartha (2009): *Inside the Integrated Assessment Models: Four Issues in Climate Economics,* Climate and Development 1(2), 166–184.

Stephan, G., Müller-Fürstenberger, G. and P. Previdoli (1997): *Overlapping Generations or Infinitely-Lived Agents: Intergenerational Altruism and the Economics of Global Warming,* Environmental and Resource Economics 10(1), 27–40.

Stern, N. H. (2008): *The Economics of Climate Change,* American Economic Review 98(2), 1–37.

Stern, N. H. (2013): *The Structure of Economic Modeling of the Potential Impacts of Climate Change: Grafting Gross Underestimation of Risk onto Already Narrow Science Models,* Journal of Economic Literature 51(3), 838–859.

Stern, N. H. (2007): *The Economics of Climate Change: The Stern Review,* Cambridge University Press, Cambridge, UK.

Stigler, G. J. and G. S. Becker (1977): *De Gustibus Non Est Disputandum,* American Economic Review 67(2), 76–90.

Tol, R. S. J. and G. W. Yohe (2006): *A Review of the Stern Review,* World Economics 7(4), 233–250.

Tvede, M. (2010): *Overlapping Generations Economies,* Palgrave McMillan, Basingstoke.

UN (2012): *World Population Prospects: The 2012 Revision*, New York. Retrieved from http://www.google.de/url?sa=t&rct=j&q=&esrc=s&source=web&cd=2&ved=0CCkQF jAB&url=http%3A%2F%2Fesa.un.org%2Fwpp%2Fdocumentation%2Fpdf%2FWPP2 012_%2520KEY%2520FINDINGS.pdf&ei=-7r0VKnXHoW3PIGTgZAI&usg=AFQjCNHlN4yBsdTgJHyUot3ZCoufCgdUHw&bv m=bv.87269000,d.ZWU Accessed on March 02, 2015.

Weitzman, M. L. (2007): *A Review of the Stern Review on the Economics of Climate Change*, Journal of Economic Literature 45(3), 703–724.

Weitzman, M. L. (2009): *On Modeling and Interpreting the Economics of Catastrophic Climate Change*, The Review of Economics and Statistics 91(1), 1–19.

Weyant, J., Davidson, O., Dowlabathi, H., Edmonds, J., Grubb, M., Parson, E. A., Richels, R., Rotmans, J., Shukla, P. R. and R. S. J. Tol (1996): *Integrated Assessment of Climate Change: An Overview and Comparison of Approaches and Results*, Cambridge University Press, Cambridge, UK.

Appendix A – Figures

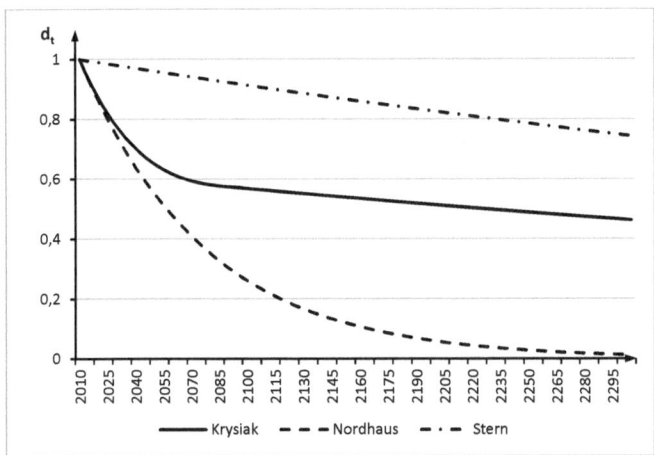

Figure 1. The discounting scheme proposed by Krysiak in comparison to a utility discount rate of 0.1 percent (Stern) and 1.5 percent (Nordhaus). The scheme is calculated for a life duration of $\hat{T} = 75$, a risk of extinction of $\theta = 0.1$ percent and a pure rate of time preference of $\rho = 1.5$ percent.

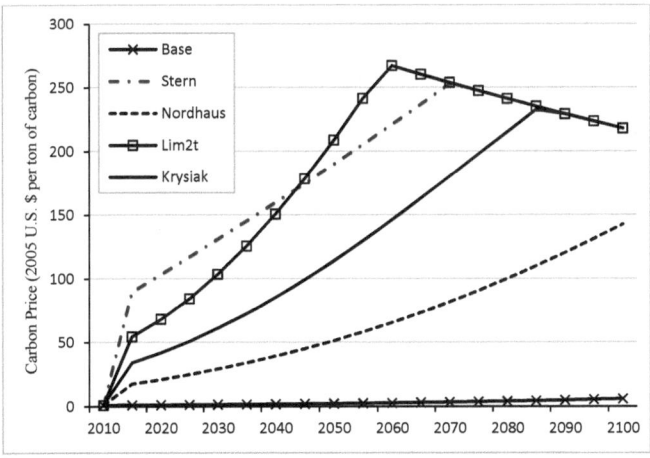

Figure 2-1. Globally averaged carbon prices under alternative policies. Prices are per ton of carbon.

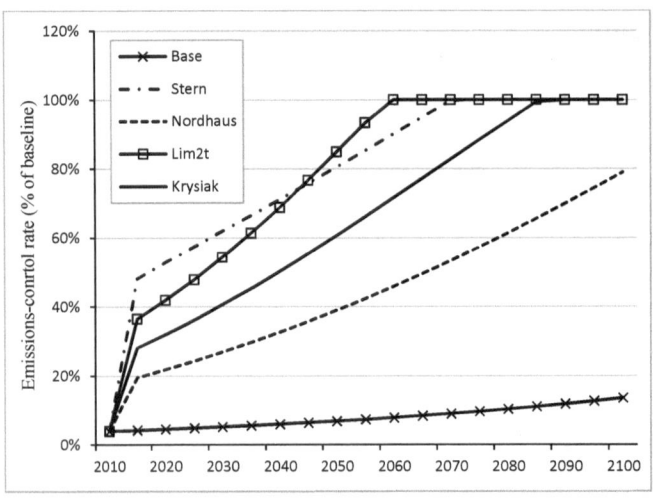

Figure 2-2. Emissions-control rates for alternative scenarios.

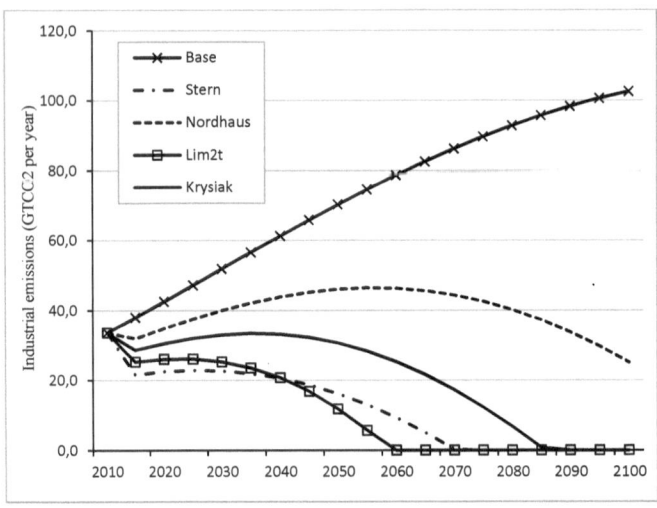

Figure 2-3. Global industrial emissions of CO_2 under alternative policies.

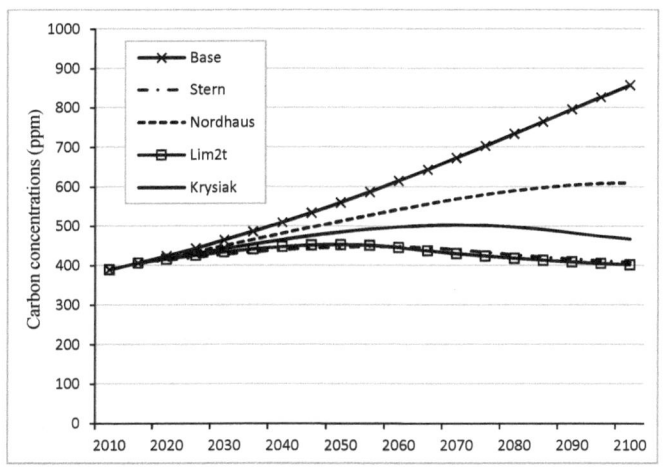

Figure 2-4. Atmospheric concentrations of CO_2 for alternative runs.

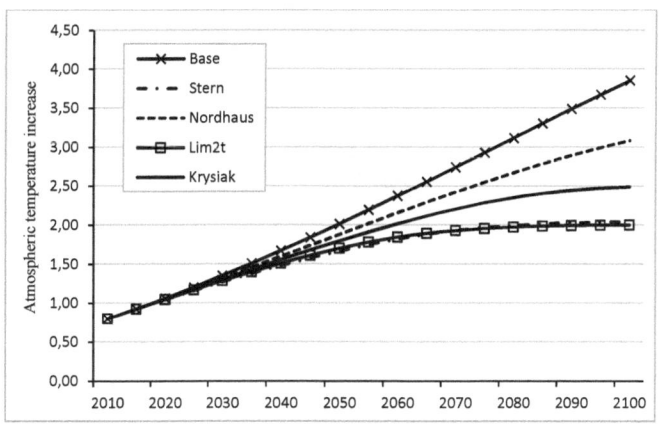

Figure 2-5. Global temperature increase in °C from 1990 under alternative policies.

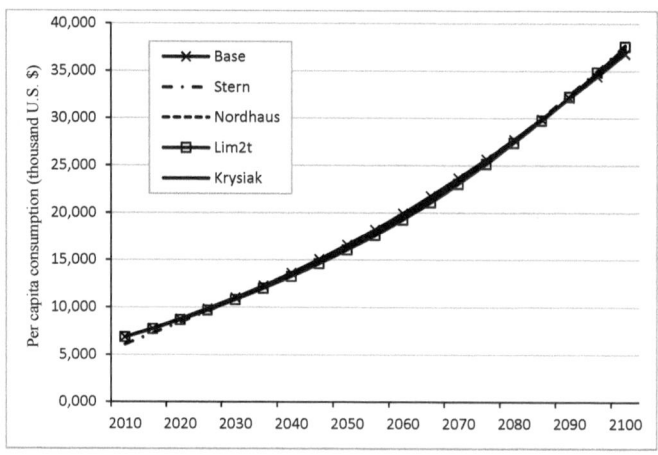

Figure 2-6. Per capita consumption under alternative policies.

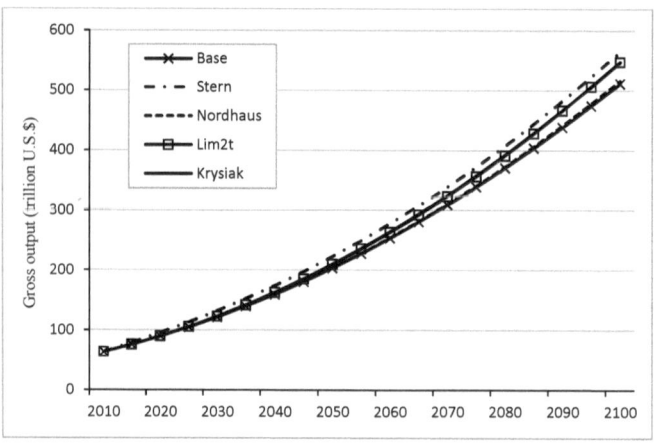

Figure 2-7. Global output under alternative policies.

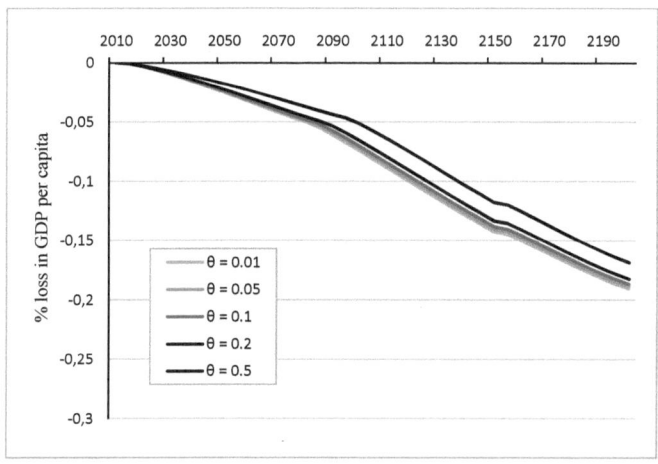

Figure 3-1. Percentage loss in per capita GDP for different values of the risk of extinction θ.

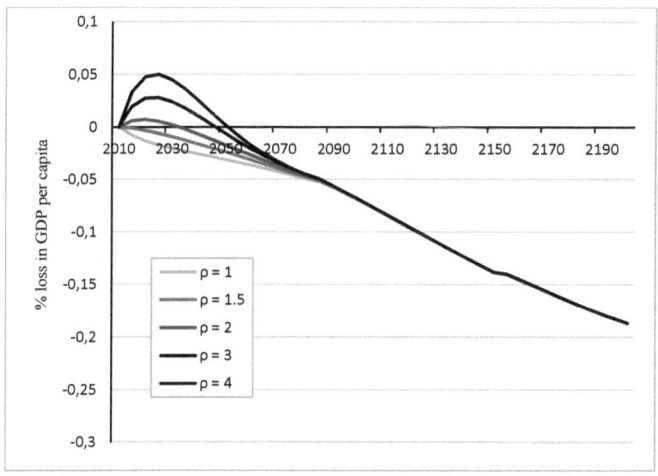

Figure 3-2. Percentage loss in per capita GDP for different values of the pure rate of time preference ρ.

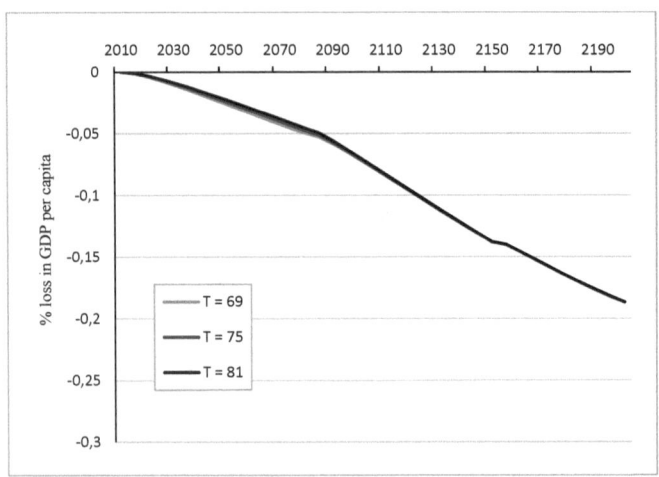

Figure 3-3. Percentage loss in per capita GDP for different values of the expected life duration $\hat{T}$.

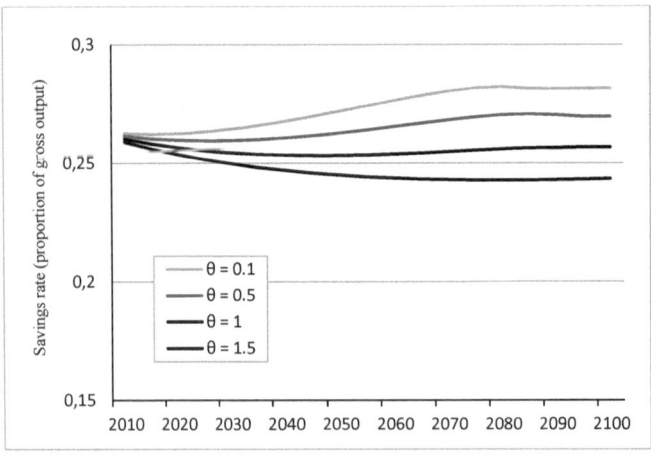

Figure 3-4. Optimal savings rates for different values of the risk of extinction θ.

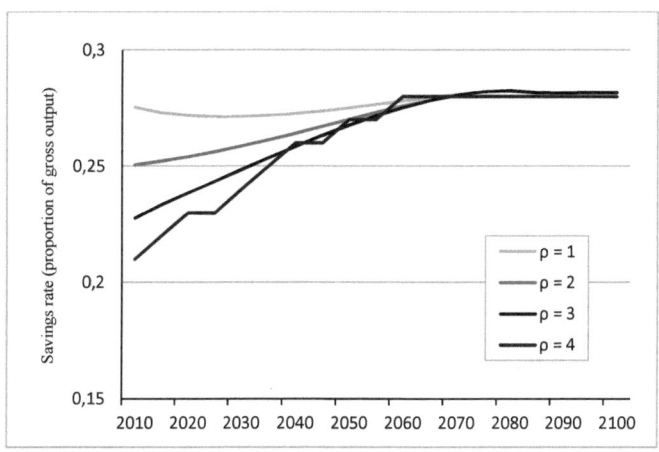

Figure 3-5. Optimal savings rates for different values of the pure rate of time preference ρ.

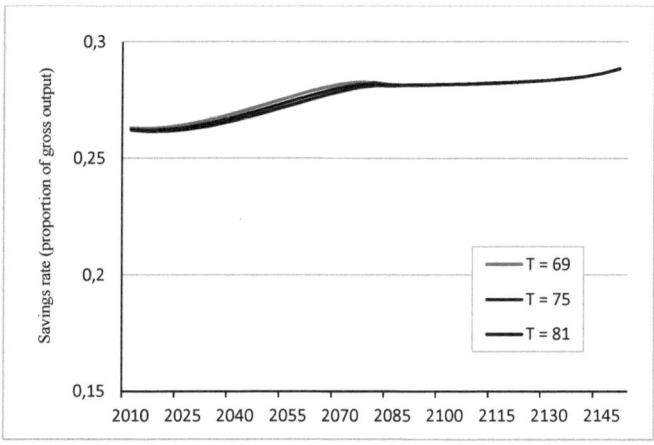

Figure 3-6. Optimal savings rates for different values of the expected life duration $\hat{T}$.

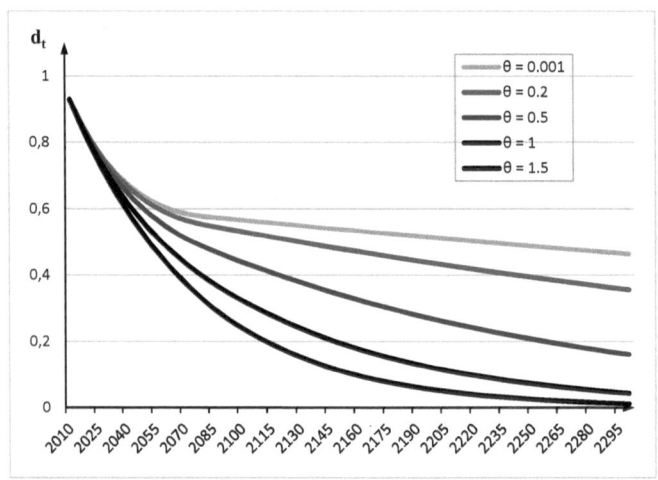

Figure 4-1. Discounting scheme proposed by Krysiak (2010) for different values of the risk of extinction θ.

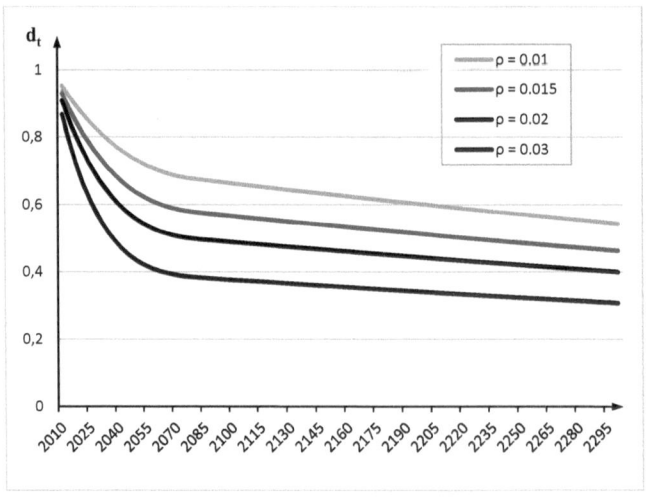

Figure 4-2. Discounting scheme proposed by Krysiak (2010) for different values of the pure rate of time preference ρ.

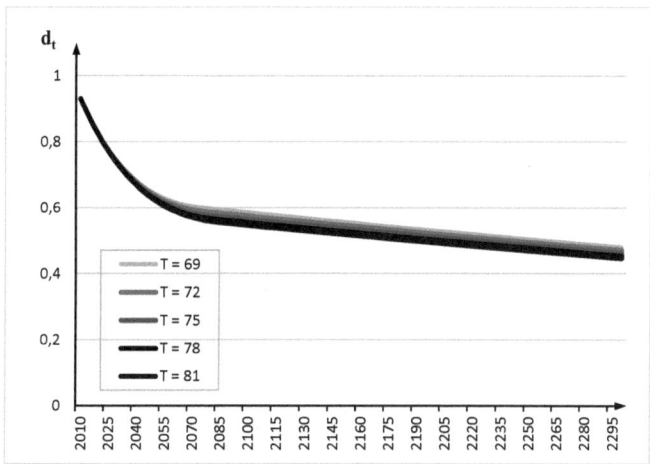

Figure 4-3. Discounting scheme proposed by Krysiak (2010) for different values of the expected life duration $\hat{T}$.

Appendix B – Tables

Table 2-1. Carbon Price (2005 $ per ton CO_2)

Policy	2010	2020	2030	2050	2075	2100	2150
Baseline	1,0	1,1	1,2	1,3	1,5	1,6	1,8
Nordhaus	1,0	21,2	29,3	51,5	90,8	142,8	169,3
Stern	1,0	89,8	103,7	117,4	131,3	145,4	159,9
Krysiak	1,0	42,0	61,3	114,1	197,8	218,1	169,3
Lim2t	1,0	68,3	103,8	208,9	247,5	218,1	169,3

Note: All prices are averaged. Carbon prices are harmonized across regions through uniform taxes or trading.

Table 2-2. Emissions-Control Rates (%)

Policy	2010	2020	2030	2050	2075	2100	2150
Baseline	4%	4%	5%	7%	10%	14%	27%
Nordhaus	4%	22%	27%	39%	57%	79%	100%
Stern	4%	53%	62%	80%	100%	100%	100%
Krysiak	4%	32%	41%	61%	88%	100%	100%
Lim2t	4%	42%	54%	85%	100%	100%	100%

Table 2-3. Industrial Emissions (GTCO2 per year)

Policy	2010	2020	2030	2050	2075	2100	2150
Baseline	33,6	42,5	51,9	70,2	89,6	102,5	103,0
Nordhaus	33,6	34,8	40,0	46,1	42,5	25,1	0,0
Stern	33,6	22,5	22,7	16,1	0,0	0,0	0,0
Krysiak	33,6	30,5	32,9	30,6	12,3	0,0	0,0
Lim2t	33,6	26,0	25,2	11,8	0,0	0,0	0,0

Table 2-4. Atmospheric Concentrations (ppm)

Policy	2010	2020	2030	2050	2075	2100	2150
Base	390	425	464	560	704	858	1134
Nordhaus	390	421	451	513	581	610	535
Stern	390	414	428	447	434	408	388
Krysiak	390	419	443	485	503	467	427
Lim2t	390	416	434	453	424	402	384

Table 2-5. Atmospheric Temperature (°C above preindustrial)

Policy	2010	2020	2030	2050	2075	2100	2150
Baseline	0,80	1,06	1,35	2,01	2,93	3,85	5,36
Nordhaus	0,80	1,05	1,32	1,88	2,55	3,09	3,30
Stern	0,80	1,05	1,27	1,67	1,97	2,04	1,96
Krysiak	0,80	1,05	1,31	1,80	2,29	2,49	2,39
Lim2t	0,80	1,05	1,29	1,70	1,96	2,00	1,92

Note: Increases are relative to the 1900 average.

Table 2-6. Consumption Per Capita (thousand USD per year)

Policy	2010	2020	2030	2050	2075	2100	2150
Baseline	6,886	8,768	11,011	16,600	25,654	36,819	64,123
Nordhaus	6,878	8,756	10,992	16,567	25,649	37,063	67,609
Stern	6,103	8,432	10,812	16,361	25,466	37,740	70,284
Krysiak	6,845	8,680	10,875	16,349	25,335	37,323	69,569
Lim2t	6,853	8,653	10,794	16,051	25,139	37,580	70,185

Table 2-7. Output (Net of Damages and Abatement, trillion USD per year)

Policy	2010	2020	2030	2050	2075	2100	2150
Baseline	63,47	89,32	120,87	200,99	331,42	491,23	878,22
Nordhaus	63,47	89,32	120,89	201,22	333,13	498,45	937,44
Stern	63,47	94,65	129,93	216,59	359,04	548,62	1049,86
Krysiak	63,47	89,60	121,98	205,90	346,70	528,33	1018,98
Lim2t	63,47	89,25	121,17	202,98	343,61	532,11	1026,90

Appendix C – Derivations

(I) Ramsey formula[89]

Suppose the utility function of an infinitely-lived individual is given by the CIES-function

$$(1) \quad u(c_t) = \frac{c_t^{1-\eta}}{1-\eta}, \text{ with } c_t, \eta > 0.$$

The utility function is a concave function of consumption c_t with diminishing marginal utility of consumption $\left(\frac{du(c_t)}{dc_t} > 0; \frac{d^2u(c_t)}{dc_t^2} < 0\right)$. η denotes the inverse of the intertemporal elasticity of substitution and is defined as $\eta(c_t) = \frac{u''(c_t)}{u'(c_t)} c_t$. The individual's present utility of her entire consumption path $(c_1, c_2, c_3, \dots)$ is given by

$$(2) \quad U(c_1, c_2, c_3, \dots) = \int_0^\infty u(c_t)e^{-\rho t}\, dt$$

where ρ reflects the individual's degree of impatience. The individual starts life with the initial capital stock k_0 and devotes her entire time to work. The capital stock in t is given by the sum of the inherited capital stock from period $t - 1$ and the unspent income. The production function is given by the function $y_t = f(k_{t-1})$ which gives the amount of production at each date t if the capital stock inherited from the last date is k_{t-1}. The production function is increasing in k_{t-1} but shows diminishing returns. The individual maximizes her lifetime utility for the given set of constraints:

$$(3) \quad \max_c \sum_{t=1}^{\infty} u(c_t)e^{-\rho t} dt$$

$s.t.:$

$$(4) \quad c_t + i_t = f(k_{t-1}), \qquad t = 1,2,3 \dots$$

$$(5) \quad k_t = (1 - d)k_{t-1} + i_t, \qquad t = 1,2,3 \dots$$

where i_t is the amount of output she chooses to invest in the capital stock in period t. Constraints (4) say that the amount of consumption and investment is limited to the output produced in each period. Constraints (5) say that the capital stock in each period

---

[89] Our derivation is based on the approach of Roemer (2011).

is made up of the investment made in the same period and the depreciated capital stock of the period before, where d gives the rate at which capital depreciates. Thus, the individual's optimization problem can be transformed in the following Lagrangian:

$$L(c_t, k_{t-1}, i_t, \lambda_t, \gamma_t) = \sum_{t=1}^{\infty} \left(\frac{1}{1+\rho}\right)^{t-1} \frac{c_t^{1-\eta}}{1-\eta} + \lambda_t[f(k_{t-1}) - c_t - i_t] +$$
$$\gamma_t[(1-d)k_{t-1} + i_t - k_t].$$

Taking the first derivative with respect to the consumption c_t, we get

$$(6) \quad \frac{\partial L(c_t, k_{t-1}, i_t, \lambda_t, \gamma_t)}{\partial c_t} = \sum_{t=1}^{\infty} -\left(\frac{1}{1+\rho}\right)^{t-1} c_t^{-\eta} - \lambda_t = 0.$$

Thus, for any two points of time $t = i$ and $t = i+1$, we can write:

For $t = i$:

$$(7) \quad -\left(\frac{1}{1+\rho}\right)^{i-1} c_i^{-\eta} - \lambda_i = 0$$

For $t = i+1$:

$$(8) \quad -\left(\frac{1}{1+\rho}\right)^{i-2} c_{i-1}^{-\eta} - \lambda_{i-1} = 0$$

Dividing (8) by (7) gives

$$(9) \quad \frac{\lambda_{i-1}}{\lambda_i} = \left(\frac{c_i}{c_{i-1}}\right)^{\eta} (1+\rho).$$

Taking the first derivative of the Lagrangian with respect to the investment i_t and the capital stock k_t, we obtain

$$(10) \quad \frac{\partial L(c_t, k_{t-1}, i_t, \lambda_t, \gamma_t)}{\partial i_t} = -\lambda_t + \gamma_t = 0 \Leftrightarrow \lambda_t = \gamma_t$$

and

$$(11) \quad \frac{\partial L(c_t, k_{t-1}, i_t, \lambda_t, \gamma_t)}{\partial k_t} = -\gamma_{t-1} + \lambda_t \frac{\partial f(k_t)}{\partial k_t} + \gamma_t(1-d) = 0.$$

For $t = i$, we can rewrite (10) and (11) as

$$(12) \quad \lambda_i = \gamma_i$$

and

$$(13) \quad -\gamma_{i-1} + \lambda_i \frac{\partial f(k_i)}{\partial k_i} + \gamma_i(1-d) = 0.$$

60

Plugging (12) in (13) yields (14):

$$-\lambda_{i-1} + \lambda_i \frac{\partial f(k_i)}{\partial k_i} + \lambda_i(1-d) = 0$$

$$\Leftrightarrow \quad \frac{\lambda_{i-1}}{\lambda_i} = \frac{\partial f(k_i)}{\partial k_i} + (1-d). \ (14)$$

Setting (9) and (14) equal it follows

$$(15) \quad \frac{\partial f(k_i)}{\partial k_i} + (1-d) = \left(\frac{c_i}{c_{i-1}}\right)^\eta (1+\rho).$$

Writing for $\frac{c_i}{c_{i-1}} = 1 + g(t)$ and defining the marginal productivity of capital net of the depreciation rate at date t as $r(t) = \frac{\partial f(k_t)}{\partial k_t} - d$ gives us

$$(16) \quad r(t) + 1 = (1+\rho)(1+g(t))^\eta$$

for all dates t. Next, taking the logarithm of (16), we obtain

$$(17) \quad \log(r(t) + 1) = \log(1+\rho) + \eta \log(1+g(t)).$$

For very small numbers of $r(t)$, ρ and $g(t)$ we can write:

$$(18) \quad r_t \approx \rho - \eta g_t.$$

Equation (18) is the so-called Ramsey rule, which was first derived in Ramsey (1928).

(II) Discounting scheme

Equation (7)

$$W = \sum_{i=-\hat{T}}^{T} \sum_{t=Max\{i,0\}}^{Min\{\hat{T}+i,T\}} \frac{g}{(1+\rho)^{t-Max\{i,0\}}} (1-\theta)^{Max\{i,0\}} U(c_t).$$

The discounting scheme divides the discounting motive in two parts: The first part (*Term 1*) includes all generations that live with certainty. These are the ones that are born before and in $i = 0$. The second part (*Term 2*) includes all generations that live with uncertainty. These are ones that are born after $i = 0$.

61

$$W = \sum_{i=-\hat{T}}^{T} \sum_{t=Max\{i,0\}}^{Min\{\hat{T}+i,T\}} \frac{g}{(1+\rho)^{t-Max\{i,0\}}} (1-\theta)^{Max\{i,0\}} U(c_t)$$

$$= \sum_{i=-\hat{T}}^{0} \sum_{t=Max\{i,0\}}^{Min\{\hat{T}+i,T\}} \frac{g}{(1+\rho)^{t-Max\{i,0\}}} (1-\theta)^{Max\{i,0\}} U(c_t)$$

$$+ \sum_{i=1}^{T} \sum_{t=Max\{i,0\}}^{Min\{\hat{T}+i,T\}} \frac{g}{(1+\rho)^{t-Max\{i,0\}}} (1-\theta)^{Max\{i,0\}} U(c_t)$$

To obtain *Term 1*, we must resort the double sum as follows:

$$\sum_{i=-\hat{T}}^{0} \sum_{t=Max\{i,0\}}^{Min\{\hat{T}+i,T\}} \frac{g}{(1+\rho)^{t-Max\{i,0\}}} (1-\theta)^{Max\{i,0\}} U(c_t)$$

$$= \sum_{i=-\hat{T}}^{0} \sum_{t=Max\{i,0\}}^{\hat{T}+i} \frac{g}{(1+\rho)^{t-Max\{i,0\}}} (1-\theta)^{Max\{i,0\}} U(c_t)$$

$$= \sum_{t=0}^{\hat{T}-\hat{T}} \frac{g}{(1+\rho)^{t-Max\{-\hat{T},0\}}} (1-\theta)^{Max\{-\hat{T},0\}} U(c_t)$$

$$+ \sum_{t=0}^{\hat{T}-\hat{T}+1} \frac{g}{(1+\rho)^{t-Max\{-\hat{T}+1,0\}}} (1-\theta)^{Max\{-\hat{T}+1,0\}} U(c_t)$$

$$+ \sum_{t=0}^{\hat{T}+2} \frac{g}{(1+\rho)^{t-Max\{-\hat{T}+2,0\}}} (1-\theta)^{Max\{-\hat{T}+2,0\}} U(c_t) + \cdots$$

$$+ \sum_{t=0}^{\hat{T}-\hat{T}+\hat{T}} \frac{g}{(1+\rho)^{t-Max\{-\hat{T}+\hat{T},0\}}} (1-\theta)^{Max\{-\hat{T}+\hat{T},0\}} U(c_t)$$

$$= \sum_{t=0}^{0} \frac{g}{(1+\rho)^{t}} U(c_t) + \sum_{t=0}^{1} \frac{g}{(1+\rho)^{t}} U(c_t) + \sum_{t=0}^{2} \frac{g}{(1+\rho)^{t}} U(c_t) + \cdots$$

$$+ \sum_{t=0}^{\hat{T}} \frac{g}{(1+\rho)^{t}} U(c_t) = \underbrace{\sum_{t=0}^{\hat{T}} \frac{g}{(1+\rho)^{t}} Max\{\hat{T}-t+1,0\} U(c_t).}_{Term\ 1}$$

Next, assuming that the discounting scheme is only applied for time horizons that are smaller than T, we can rewrite the double sum:

$$\sum_{i=1}^{T} \sum_{t=Max\{i,0\}}^{Min\{\hat{T}+i,T\}} \frac{g}{(1+\rho)^{t-Max\{i,0\}}} (1-\theta)^{Max\{i,0\}} U(c_t)$$

$$= \sum_{i=1}^{T} \sum_{t=i}^{\hat{T}+i} \frac{g}{(1+\rho)^{t-i}} (1-\theta)^{i} U(c_t)$$

$$= \sum_{i=Max\,(t-T,1)}^{t} \sum_{t=0}^{T} \frac{g}{(1+\rho)^{t-i}} (1-\theta)^{i} U(c_t).$$

Simplifying the outer sum yields:[90]

$$\frac{(1+\rho)^{-t} \left(\left(-(-1+\theta)(1+\rho)\right)^{Max\{t-\hat{T},1\}} - \left(-(-1+\theta)(1+\rho)\right)^{1+t} \right)}{\theta + (-1+\theta)\rho}$$

$$= \frac{(1+\rho)^{-t} \left(-\left((1-\theta)(1+\rho)\right)^{1+t} + \left((1-\theta)(1+\rho)\right)^{Max\{t-\hat{T},1\}} \right)}{\theta + (-1+\theta)\rho}$$

$$= \frac{(1+\rho)^{-t} \left(\left((1-\theta)(1+\rho)\right)^{1+t} - \left((1-\theta)(1+\rho)\right)^{Max\{t-\hat{T},1\}} \right)}{(1-\theta)\rho - \theta}$$

$$= \frac{(1+\rho)(1-\theta)^{1+t} - (1+\rho)^{-t}\left((1-\theta)(1+\rho)\right)^{Max\{t-\hat{T},1\}}}{(1-\theta)\rho - \theta}$$

$$= \frac{(1+\rho)(1-\theta)^{1+t} - (1+\rho)^{-t}\left((1-\theta)(1+\rho)\right)^{Max\{t-\hat{T},1\}}}{(1-\theta)\rho - \theta}$$

$$= \frac{(1+\rho)(1-\theta)^{1+t} - (1+\rho)^{-t}\left((1-\theta)(1+\rho)\right)^{1+t}\left((1-\theta)(1+\rho)\right)^{Max\left\{-\hat{T}-1,-t\right\}}}{(1-\theta)\rho - \theta}$$

$$= \frac{(1+\rho)(1-\theta)^{1+t} - (1+\rho)^{-t}\left((1-\theta)(1+\rho)\right)^{1+t}\left((1-\theta)(1+\rho)\right)^{-Min\{\hat{T}+1,t\}}}{(1-\theta)\rho - \theta}$$

$$= \frac{(1+\rho)(1-\theta)^{1+t} \left(1 - \left((1-\theta)(1+\rho)\right)^{-Min\{\hat{T}+1,t\}}\right)}{(1-\theta)\rho - \theta}.$$

---

[90] See the enclosed Mathematica file.

Thus, we can write

$$\sum_{i=1}^{T} \sum_{t=Max\{i,0\}}^{Min\{\hat{T}+i,T\}} \frac{g}{(1+\rho)^{t-Max\{i,0\}}} (1-\theta)^{Max\{i,0\}} U(c_t)$$

$$= \sum_{t=0}^{T} g \underbrace{\frac{(1+\rho)(1-\theta)^{1+t} \left(1 - \left((1-\theta)(1+\rho)\right)^{-Min\{\hat{T}+1,t\}}\right)}{(1-\theta)\rho - \theta}}_{Term\ 2} U(c_t).$$

Combining *Term 1* and *Term 2* we get

$$W = \sum_{t=0}^{T} \left(\frac{g}{(1+\rho)^t} Max\{\hat{T} - t + 1, 0\} \right.$$

$$\left. + g(1-\theta)^t \frac{(1-\theta)(1+\rho)}{(1-\theta)\rho - \theta} \left(1 - \frac{1}{((1-\theta)(1+\rho))^{Min\{t,\hat{T}+1\}}}\right) \right) U(c_t).$$

Last, normalizing the discounting scheme with the weight $\frac{1}{\hat{T}+1}$ we obtain Equation (8):

Equation (8)

$$W = \sum_{t=0}^{T} \frac{1}{\hat{T}+1} \left(\frac{g}{(1+\rho)^t} Max\{\hat{T} - t + 1, 0\} \right.$$

$$\left. + g(1-\theta)^t \frac{(1-\theta)(1+\rho)}{(1-\theta)\rho - \theta} \left(1 - \frac{1}{((1-\theta)(1+\rho))^{Min\{t,\hat{T}+1\}}}\right) \right) U(c_t).$$